Homemade Lightning

Creative Experiments in Electricity

R. A. Ford

TAB BOOKS
Blue Ridge Summit, PA

MW01010503

FIRST EDITION
SECOND PRINTING

© 1991 by **R. A. Ford**.
Published by TAB Books.
TAB Books is a division of McGraw-Hill, Inc.

Printed in the United States of America. All rights reserved. The publisher takes no responsibility for the use of any of the materials or methods described in this book, nor for the products thereof.

Library of Congress Cataloging-in-Publication Data

Ford, R. A.
Homemade lightning : creative experiments in electricity / by R.A. Ford.
p. cm.
Includes index.
ISBN 0-8306-7576-0 ISBN 0-8306-3576-9 (pbk.)
1. Electrostatic apparatus and appliances—Experiments.
2. Electric generators—Experiments. I. Title.
QC573.F67 1990
621.31′3—dc20 90-44414
CIP

TAB Books offers software for sale. For information and a catalog, please contact TAB Software Department, Blue Ridge Summit, PA 17294-0850.

Acquisitions Editor: Roland S. Phelps
Technical Editor: Andrew Yoder
Production: Katherine G. Brown
Book Design: Jaclyn J. Boone
Typesetting: Terry Hite

Contents

Acknowledgments *vi*
Introduction *vii*
Lightning/Electrocution Safety *ix*

PART 1: DESIGN AND CONSTRUCTION

1 Types of electrostatic generators *3*

2 Elements of good design *5*
Winter's electrical machine *5*

3 James Wimshurst's influence machine *11*
Wimshurst machine *15*
Modified Wimshurst machines *16*

4 The author's generator *19*
The stand *21*
Support shaft for discs *22*
Bosses for the discs *22*
Drive belts and pulleys *24*
Belt tension and motor mount *26*
Charge collector support system *26*
Charge collector combs *28*
Discharge terminals, supports, and rods *34*
Neutralizer rods and blades *39*
12-volt power supply *41*
Charging the generator *43*
Leyden jar condenser *46*

5 Unusual generator designs *53*
Experimental design modifications *59*
Experimental disc materials *59*
Precision construction techniques *67*
Liquid, gas, and vacuum chambers *68*

6 Theories of generator operation *69*

PART 2: ACCESSORY INSTRUMENTS, EXPERIMENTS, AND APPLICATIONS

7 The electroscope *77*
Building a cookie-tin electroscope *81*
Directions for building the cookie-tin electroscope *81*
Electroscope anomalies *87*
Research avenues *89*

8 The Leyden jar condenser *93*
Dielectric constants *94*
Design modifications of Leyden jars *96*

9 The electrophorus *101*
The grounded electrophorus *103*
Electrical shock from a piece of paper *103*
Making the traditional cake electrophorus *104*
- Baking a cake *104*
- The author's electrophorus *106*

Semiconductive stones *112*
Some avenues for electrical pioneers *114*
Summary *114*

10 Electrostatic motors *115*

11 Electrohorticulture *123*

12 Electrotherapeutics *127*

13 High-voltage humans *133*

14 Cold light *137*
The auroral light *137*
Earthquake lights *140*
Invisible phosphorescence *141*
Phosphorescent lamps *143*

15 Miscellaneous experiments ***145***
The levitating rocket *146*
Modified levitating rocket *147*

16 Electroaerodynamics ***151***

17 Countergravitation ***153***
Kinetic gravitational theory *154*
Dr. Nipher's deflection experiments *156*

18 Unusual electric discharges ***163***
Lightning shadowgraphs *163*
Tornadoes as electrical machines *163*
The electrical entities *169*
Historic entity experiments *170*
Theoretical implications *176*

19 Some philosophical conclusions and insights ***179***
Visualization *179*
Intuition *180*
Qualitative anatomization *180*
Some useful surroundings *182*

Appendix ***184***

Bibliography ***187***

Index ***195***

For Timothy Paul

Acknowledgments

SPECIAL THANKS TO: A. D. MOORE WHOSE BOOKS GOT ME INTERESTED IN electrostatics; to Franklin B. Lee who has provided an affordable, high-quality electrical apparatus to the public over a period of many years; to Diana for her considerable bravery in modeling for photos; to Laura for her assistance in taking journal photos; to Charles for his machinist skills; and to the library staff at the University of Illinois, Urbana and the interlibrary loan services of the Lincoln Trails Library System.

Introduction

YEARS AGO, DURING MY LAST YEAR IN HIGH SCHOOL, I DECIDED TO SAVE up and buy a high-voltage Van de Graaff generator in kit form. I assembled and experimented with this machine, which had a spark potential of 500,000 volts; in the dry Connecticut winters it produced a 17-inch spark every second between its two large terminals. Many weekends were spent experimenting with these strange electrical forces and building capacitors for hotter sparks.

The fascination with high-voltage work in the lab and outdoors in nature's magnificant displays was kindled in me back then and has grown over the years. Even at that early stage, I knew this was an area of research that was not well understood or explained, and I wondered whether such generators might ever have real, practical uses. Of course, today there are many industrial applications for electrostatics, such as removing dust from smoke stacks, paint spraying, and photocopying. But could an entirely new technology be developed for exploring the forces of nature and be as practical as was the application of steam power?

Modern electrostatics began about the year 1660, when Otto von Guericke built spinning sulphur spheres, which were charged by the friction of the hand. Considerable experimenting was done from the latter half of the 1700s to about 1900. By this time, increasing interest was developing in the applications of that other aspect of electrical science—electromagnetism. Michael Faraday, Nikola Tesla, and Thomas Edison were largely responsible for this development. In time, physics books would treat the study of electrostatics primarily as an entertaining novelty with no practical use.

In spite of the centuries of work, we still need much more understanding of the nature of electric "charge", of how electric forces act across space, and of how electric potential energy is stored. At the very least, this would help to understand how thunderstorms develop.

This book is divided into two parts; the first part describes high voltage generator design and construction with a brief mention of theory of operation. The second part details the basic instruments used with the generator and some of the areas wide open for pioneers. As the title implies, this book is not written for the couch potato who waits passively to be entertained, but rather for those who love to build, experiment and investigate along original lines of research. I am especially interested in encouraging high school students and their teachers who want to do something really unusual for their science fair projects. A basic grasp of electrostatic principles is helpful on the theoretical side.

The Wimshurst disc-type generator and its modifications is selected as a very versatile, compact, and reliable machine. An intermediate level of skill in wood, metal, and plastics working, as given in high school industrial arts classes, is needed for generator construction. In this way, hands-on physics experiments are combined with manual skills that emphasize the importance of the industrial arts, directed towards the high goal of scientific research. Sources for construction materials are provided.

To those who work in the high-technology field but find their job uninspiring—who need more feel for practice than for pure theory or find themselves morally opposed to the direction their technical work is taking—take heart. There is plenty of room in the science of electrostatics for creative thinking!

In order to get as many readers involved as possible, I have designed my generators and accessories with the idea of reasonable cost and readily available materials foremost in thought. I provide sources for parts not locally available.

When you are finished with this book, you will be able to walk into hardware, fabric, auto parts, and arts and crafts stores and see completely novel uses for ordinary products. Of course, the store clerks might be baffled by your excitement!

A well-made high-speed generator can deliver much more voltage at reasonable charging current as industrial power supplies costing thousands of dollars, yet without transformers, rectifiers, and filter systems. Although high-voltage electricity does have great entertainment value, my main objective is to revive it as a serious tool for physics research.

Whatever your interest, please enjoy and happy innovating!

Lightning/ Electrocution Safety

ALTHOUGH LIGHTNING FEELS NO COMPULSION TO OBEY OHM'S LAW AND IT remains unpredictable, a few commonsense rules will reduce the hazards.

- Stay away from large bodies of water and wide-open spaces.
- When indoors, stay away from windows and fireplaces. Never talk on the telephone during violent storms.
- Stay in the car if you are isolated. All-metal cars, trains, and airplanes are fairly safe. Fiberglass or composite-fiber bodies will require a special means for diverting a strike.

In the case of electrocution, the victim should be treated as anyone needing CPR would be. Should this fail to get results, all is not lost as is illustrated by this case, mentioned many years ago in a science journal. A workman was electrocuted when he contacted a high-voltage power line. After all efforts to revive him failed, a most novel idea was hatched. As the victim lay on his back, one person removed his shoes and socks, and lifted up his legs, and held them together. A second person, with a large flat paddle, administered sharp blows to the soles of the workman's feet. Shortly, he was aroused and recovered fully—sort of like spanking the newborn baby, I suppose! I will leave it to experts to explain how this peculiar method of resuscitation worked.

Remember, the victim of electrocution might appear clinically dead, yet still not be beyond hope.

1
Part
Design and Construction

1

Types of electrostatic generators

THERE ARE PRESENTLY TWO MAIN CLASSES OF ELECTROSTATIC GENERATORS. The first type charges by *frictional slippage*, or *impact*. This means there is direct physical contact between two different material surfaces. Frictional charging in an earlier time was called *triboelectricity*, *tribo* being the Greek term meaning "to rub." Otto von Guericke's spinning sulfur sphere (1663) which rubbed against the hand, was the early form. Later in 1768, Mr. Jesse Ramsden and Mr. Jan Ingenhousz developed glass disc generators which rubbed against a leather pad coated with metallic powder instead of the hand. The frictional generator using glass discs reached a high level of development in 1856 with Karl Winter's design. For a good description of frictional generators, see the book *Early Electrical Machines* by Bern Dibner.

The second class of machines are called *electrostatic induction generators*. The word *induction*, according to the *Oxford English Dictionary*, in this context means to bring about an electrified state in a body by *proximity*—closeness without contact—of another electrified body. The induction generator was originally called an *influence machine*; the change in names occurred gradually during the years 1890 to 1920. I prefer the word *influence* because it gives a clearer mental picture. Oxford defines *influence* as "the action or inflow of immaterial things, the operation or infusion of which is unseen or insensible—that is, only the **effects** are visible. The word *induction* draws a blank in the mind because it avoids all mention of the operating causes. I will discuss this theoretical subject later in detail. It is at the heart of a better understanding of just how electrification occurs, and especially the nature and properties of space itself. But first, I will describe the generators, which are our tools, as well as the novelties of their design and construction.

2
Elements of good design

AFTER TWO YEARS OF INTENSIVE RESEARCH ON BOTH FRICTIONAL AND and influence generators dating from the late 1700s through the 1920s, I've concluded that one of the best machines, as far as general design is concerned, was the one that appeared circa 1856 in Vienna, Austria. The inventor, electrical genius Karl Winter, put together the best of his design innovations. So popular and effective were his machines that they were still in use by electrotherapeutics practitioners up to 1930! After having tried out several different generator designs, I find myself inevitably returning to his basic design elements, which can be applied to any high-voltage disc or drum generator.

Shown in Fig. 2-1, Winter's machine was greatly popularized throughout Europe after its introduction into Edinburgh, Scotland, by Dr. Robert M. Ferguson, who also mentioned it in his books on natural philosophy (physics). A rare description appears in the *English Mechanic*, which I have rearranged.

Winter's electrical machine*

In number 27, we expressed our determination to close our columns for the present to the subject of electricity, but the flood of letters which have come to hand has induced us to keep open for a week or

*From the English Mechanic, October 20, 1865

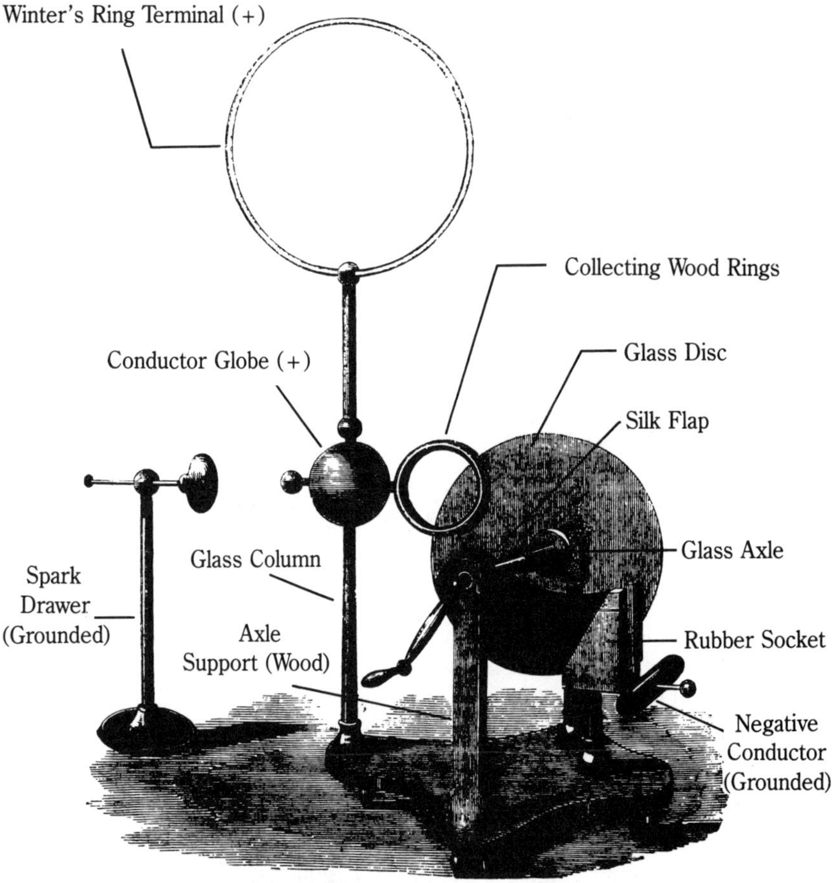

2-1 Karl Winter's machine (1856) Parts List Added. *English Mechanic,* 1865

two longer. To satisfy a crowd of correspondents, who seem altogether to have forgotten that what they have asked for has been published at length, and to enlighten several on a few new points of interest, we give an illustration and description of the machine invented by Herr Carl Winter, of Vienna.

It was first introduced into this country by Dr. R.M. Ferguson, of Edinburgh, in 1857. The Doctor had seen one in Vienna and different cities in Germany before he had one made. The largest he saw was in the Vienna Polytechnic School; it had a plate 5 feet in diameter which gave off sparks 4 feet long. However, a very good size is a 2 feet plate. At first the pillars were made of glass, but Mr. Varley, of London, made his large-sized one of vulcanite, and found the insulation improved.

The machine is essentially a plate-friction machine, and with a 2 feet plate will give 11 inch sparks. If made in the ordinary way, such a machine would be considered very successful if it gave, in the most favourable circumstances, a spark of 5 or at most 6 inches. All who have had to do with electrical machines know how extremely difficult it is, even with the requisite means of drying, to keep them in a state favourable to ready action for any length of time; though Winter's machine he left in a room for months where, during that time, there has been no fire, yet by the first turn of the plate it will give sparks 10 inches long. The cause of this efficiency may principally be attributed to the rubbers he employs, the perfection of insulation, and to the contrivance for lengthening the spark. The accompanying illustration represents one of these instruments.

It will be seen that the plate is fixed into an axle which revolves in two upright supports. One of these, in which the shorter wooden end of the axle revolves is made of glass, and the other, in which the longer glass end of the axle revolves, is of wood. Thus the electricity formed upon the plate cannot reach the ground on either side, for on the one side the insulating glass pillar, and on the other the insulating glass axle, prevents it.

The friction, as usual, is caused by pressing on the plate of glass a flat surface of leather covered with an amalgam of mercury, zinc, and tin, which is put on with the aid of a little grease. The frame standing on the low glass support to the right of the figure is the wooden rubber frame, into the notches of which fit two flat pieces of wood covered in front or on the side next to the plate with leather, and a very little stuffing, and provided on the other side with springs, which, acting against the frame, keep the front surface uniformly pressed against the plate. There is only one pair of rubbers, not two, as in ordinary machines, and this enables them to be placed at a greater distance from the prime conductor of the machine.

The brass ball on the tall glass support to the left is the prime conductor. For more perfect insulation this ball is fitted on to the support by means of a trumpet-shaped opening made in it, thereby preventing the dispersion of electricity that would arise from the sharp edge of a hole exactly large enough for the rod. There are three other openings in this ball, one on each side, and one at the top. The two small rings which are seen projecting upon the plate fit into one of these by means of a t-shaped piece of brass. They are made of wood, and have a groove cut in them on the side turned towards the plate, into which a row of fine pin points is fixed for collecting the electricity formed upon it.

These points are connected with the prime conductor by means of a strip of tin-foil which lines the bottom of the groove. Two wings of oil silk, attached to the rubbers, stretch between them and these rings, so as to prevent the electricity from dispersing itself before reaching them. The opening in the top of the ball is made to receive

> the stalk of the wooden ring, which is seen surmounting it, and which forms the most peculiar feature of the instrument. The function performed by this remarkable appendage is to lengthen the sparks given by the machine. It is a wooden ring, and may be from 32 to 36 in. diameter, with a thin iron wire inside to form a core, which also passes down the wooden stalk, and is in metallic connection with the brass ball or prime conductor.
>
> This ring can be removed at pleasure; when removed, it is an ordinary plate machine of the best construction. The sparks then are straight and thick, and not more than two or three inches long, whenever the ring is put on they are at once lengthened to 10 or 12 inches. It would seem that the electricity accumulates with great intensity on the thin wire which forms the core above noted, the wooden covering preventing dispersion.
>
> Dr. Ferguson tried different rings; he had one of polished iron wire, and got 6 in. sparks; covered with gutta percha he obtained 12 in. sparks; puncturing the covering the length of spark fell to 6 in. Other kinds of rings were tried, but none proved so good as that described. This is probably owing to its semi-conducting power, or superior elasticity, or in that it cannot be permanently punctured like gutta percha.
>
> To the left of the figure is the spark drawer, for receiving the sparks from the machine. The length of the spark is the test of the construction of a machine, and it would appear that in this respect Winter's holds first rank. Indeed, it is something quite novel in the history of electrical apparatus; concerning which, Mr. W. Hart of No. 7, North College Street, Edinburgh, has expressed his willingness to answer correspondents by letter.

Of special interest is Winter's extensive use of wood, a semiconductor, for high-voltage charge collection. Further details were given in *Physical Technics* by Dr. J. Frick (see Bibliography).

The collecting apparatus consists of two thick, polished wooden rings, from 2 to 5 centimeters thick with an external diameter of 13 centimeters. A groove is cut on the inside of each ring—facing the glass disc. This groove is lined with tin-foil and set with a thick row of fine pin points, which reach only to the surface of the ring. The elements of good generator design used by Karl Winter include the use of large, smooth surfaces of polished wood for collecting electricity from the disc.

Many electricians in later years forgot this detail and made small wire collectors, which easily leak charges at high potential. Evidently, these later inventors had not studied or been aware of Winter's designs, the result being that the voltage or charge potential produced in those generators was inferior. By surrounding metal with wood, a transition from conductor outward through wood (semiconductor) to the surrounding air (insulator) results. This absence of an abrupt change further reduces high-voltage leakage.

I have no doubt that several of the early frictional generators could outperform many of the later influence machines. In Winter's design, spark length was four-fifths the diameter of the glass disc, which is very good! The current output, in those days, was difficult to measure.

Much of the history of these early electrical devices was lost because the letters describing them have since turned to dust. Karl Winter's ingenious design is one of the best from that era.

3

James Wimshurst's influence machine

THE BASIC INFLUENCE (INDUCTION) GENERATOR HAD ITS BEGINNINGS WITH John Canton's *theory of electrification without touching* from about 1753. He visualized an electrical *atmosphere*, a medium that surrounds electric bodies and acts across space.

Later in 1787, Abraham Bennet described a "doubler" generator, which via influence continually increased very small charges until they built up to measurable values. Throughout the 1800s, new designs appeared; their number began to accelerate about 1860.

It is not possible to describe all the designs, but the most popular were Varley's machine (1860), Toepler's machine (1865), Holtz's machine (1865), Leyser's machine (1873), and Voss' machine (1880). Of these, the Holtz generator would continue to be used for many years in electrotherapeutics and was able to outperform the Wimshurst generator in good weather.

But these earlier designs suffered from either difficulty in starting in bad weather or from polarity reversal. Reversal of polarity between the two output terminals could occur spontaneously, so constant reliable output was hampered. In 1878, James Wimshurst of England set about to remedy these problems and improve on the Holtz design. In 1883, the basic Wimshurst machine began to appear in the science journals. His original design is shown in Fig. 3-1.

As can be seen in this figure, no provision was made for storing charges. In 1882, charge storage was added by using two Leyden jars, the first capacitors. This method increased the efficiency and increased the spark length between the discharge terminal balls.

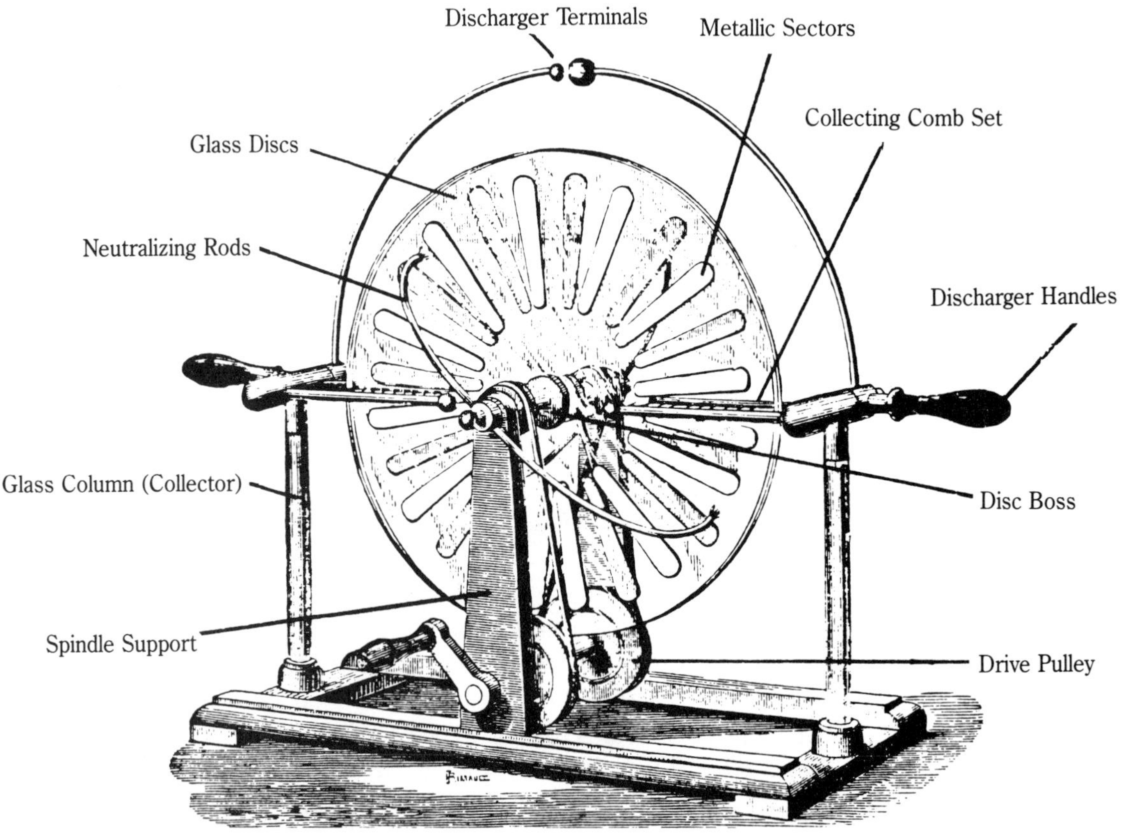

3-1 James Wimshurst's original design. *Electrical Influence Machines*, John Gray, 1903

The Wimshurst generator proved to be quite reliable in starting during the damp weather in England because of the metallic sectors on the two discs. A single turn of the handle would cause the two glass plates to literally bristle with electricity.

In addition, Wimshurst's design was not subject to polarity reversal like other machines. These two desirable features resulted in far-reaching popularity for the Wimshurst machine in Europe. Many articles appeared on variations of his basic concepts.

The largest Wimshurst was built in 1885 and had two 7-foot glass discs 3/8 inch thick, each weighing 280 pounds! (See Fig. 3-2.) When operating, this machine gave a torrent of hot sparks and electric flames up to 22 inches long.

I did manage to track this device down. As of this writing, it is located in a glass case on the second floor balcony of the Science and Industry Museum in Chicago, Illinois. I found it inoperable and in need of major repairs, but still impressive to see. Perhaps the museum staff will see fit to put it into operation for the benefit of scientific inquirers.

THE WIMSHURST 7-FT. DUPLEX ELECTRIC MACHINE.

3-2 The largest Wimshurst machine (1885). *Engineering*, Vol. 39, 1885

Another variation of the design enclosed the machine in an airtight case with chemical drying agents. Dry air helped boost voltage output and made starting easier (See Fig. 3-3). This commercial design was used by electrotherapeutic practitioners. The charge collector has been enlarged in size but is still entirely made of metal. By using multiple glass plates on a single axle, the current could be increased just as one would join batteries in a parallel circuit (The voltage remains the same.).

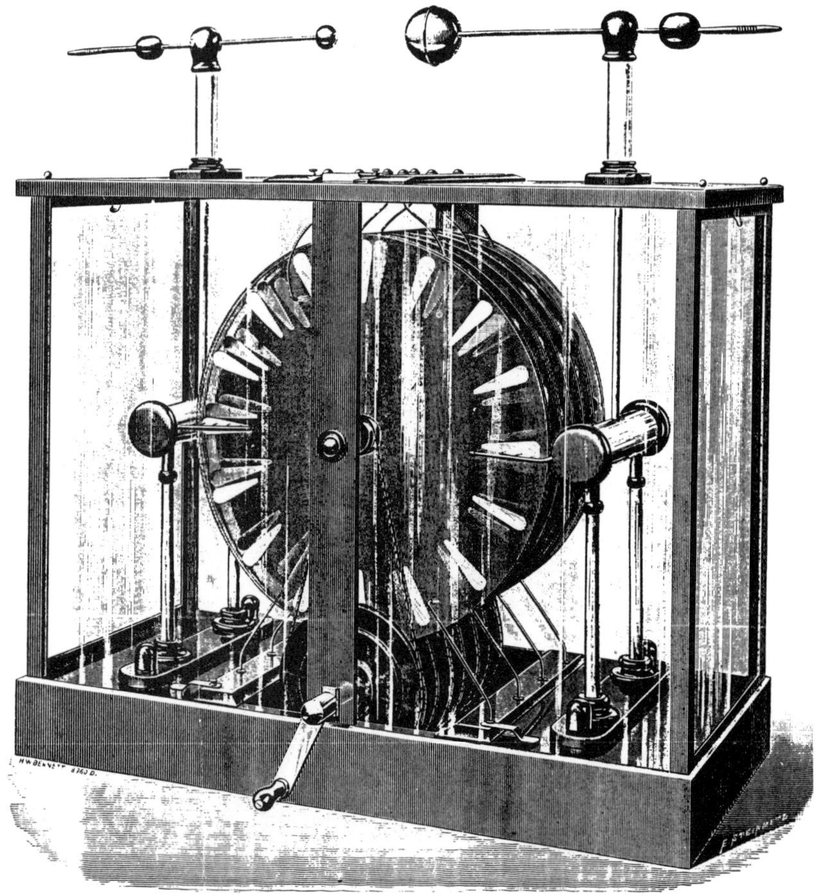

3-3 A multiplate Wimshurst (1888). *Engineering*, Vol. 45, 1888

The bottom circuit for the Leyden jar condensers use a wire to join and ground the four outer coatings of the four condensers. *Ground* here means through the woodwork, which is usually sufficient. The inner coatings of each pair of condensers is joined to its respective collecting comb. The Leyden jars should be crystal, crown, or "pyrex-type" glass so that high-voltage leakage is reduced. The collecting points can be steel dress maker's pins; the gap between pinpoints and disc surface should be greater than 1/16 inch, but not more than 1/8 inch.

Notice that the discharging balls are different sizes. This geometry results in a greater spark length. For an 18-inch diameter disc machine, Mr. Wimshurst preferred a small ball 1/2 inch in diameter and a larger ball of $1^1/_2$ inches in diameter. These two balls should be interchangeable, rather than permanently soldered on, and should depend on terminal polarity for placement.

The neutralizer brushes must make contact with the disc's metallic sectors simultaneously across the disc diameter. For this reason, only an even number of sectors can be used. Bend the arms slightly to make these contacts possible.

Brushes for the neutralizers should be springy, since they make light contact with the metallic sectors on the discs. Phosphor bronze or steel music wire 0.002 to 0.005 inch in diameter are servicable for this purpose, but they should not be so stiff that they wear away the sectors. The metallic sectors can be cut from brass or steel shim stock or adhesive-back stainless steel foil, 0.001 to 0.003 inch thick. These sectors should be cut out with sharp paper scissors, not with metal snips; snips leave a serrated edge that causes leakage. I recommend attaching sectors with epoxy glue since shellac varnish is not a good adhesive in damp weather. The sectors should be attached after the discs are varnished.

The method adopted for varnishing the glass discs, to make the surface resistant to moisture films is as follows.

Wimshurst machine*

The following is the best method for varnishing the glass discs: first mix your varnish (shellac and methylated spirits) in a bottle having a large neck; the brush ought to be more than 1 in. in diameter, and its handle should be fixed in the bung of the bottle; this arrangement keeps the varnish and the brush free from dust. Next make up some simple sort of turntable, on which you place the disc while it is being varnished. The next step is to wash the discs and dry them; then warm each disc to about the temperature of blood heat, place one on the turntable, and with the left hand set it in motion. Take the brush full of varnish from the bottle with the right hand and bring it on to the disc a little from the edge; move the brush slowly to the edge and then back, so as to finish and to lift the brush off at the centre; let the disc remain a few minutes for the varnish to set, and then finish drying in a warm place. It will then be seen the coating is so even as to be almost unnoticed. If the brush be lifted up from the disc while the varnishing is being done, the surface will be covered with air bubbles, and look unsightly. Done in this way the varnish lasts for years.

James Wimshurst

---

*From *English Mechanic*, 1889.

Warming the discs drives off moisture; the room should be dust-free and dry, of course. *Methylated spirits* is also called *denatured alcohol solvent*, and the shellac is in the form of refined flakes. If you add one or two drops of castor oil per pint of varnish, there will be less tendency for the varnish surface to crack and dry out.

The usual means for attaching the glass discs to the wood bosses is as follows:

Cut out a thin leather washer, about 1/32 to 1/16 inch with an outer diameter equal to the diameter of the boss. The hole in the washer should be a bit larger than the shaft through the discs. Cover one side of the washer with epoxy glue, or any cement for repairing glass, and press the tacky side onto the boss end. Coat the top side of the washer, and let it stand until the glue is tacky. Press the boss onto the glass disc, carefully centering, so that the boss hole is concentric with the disc diameter. Apply weight on top until the glue is dry, usually overnight.

Mr. Wimshurst also used thin leather washers cemented with bicycle tire cement. Always use washers because they cushion the discs and prevents cracking. Finally, the discs are spaced apart from each other, about 1/32 inch by using metal, fiber, or leather washers as spacers—the discs must not rub or touch at any point.

The traditional Wimshurst machine was hand-powered. Although such machines can be motor-powered for convenience, as my generator is, the speed of the glass discs must be kept low to avoid breakage.

There is a certain fascination with hand-powered versions because they produce miniature lightning bolts without any attached power cord—seemingly drawing electricity out of thin air

Modified Wimshurst machines

In spite of the glowing reports in English science journals about the performance of Wimshurst's generators, there were detractors who pointed out defects in his designs. Some of these inventors put forward their own innovations, which proved to be more efficient.

Under the best of conditions, the traditional Wimshurst could develop a spark length equal to the radius of the disc, but this was rare for most homemade units.

The current or quantity of charge was just as important as spark length. The usual method for comparing two designs of the same size and speed was to see how fast each could charge a Leyden jar to a fixed voltage. Results were not very accurate and this area of research needs to be explored. It should be clear that influence machines basically have their voltages determined by disc diameter, so doubling the diameter approximately doubles the voltage. Current output also increases with disc rotational speed, and of course with added pairs of discs on the same axle.

Methods for improving efficiency will be mentioned later. During the 1890s and onward, modified Wimshurst machines appeared. One of the important changes was to remove all sectors on the discs. These generators then became known as *sectorless Wimshurst machines*.

Most notable were the inventors Mr. Picolet (1892) and Mr. Bonetti (1893). Figure 3-4 describes Picolet's design.

3-4 A sectorless Wimshurst (1894). *Scientific American*, 1894

Sectors were found to be a cause of serious leakage, and their only advantage was to make the machine self-starting. I find that having to "charge" the sectorless generator is simple and is a good safety feature. In addition, the plates can easily be kept clean. As shown in Picolet's generator (*Scientific American*, May 1894), the neutralizer brushes are long with several points, but these points do not need to touch the disc.

Another solution to the metallic sector leakage problem was put forward in two patents by Mr. Wommelsdorf (U.S. patents 882,508 March 1908 and 1,071,196 August 1913).

He sandwiched his sectors, or *carriers*, within two thin discs. Electric contact to these carriers was through the compound disc rim only.

Even though his generator was not a true Wimshurst, the technique would still apply. Wommelsdorf claimed a large improvement in current output and a spark length of two-thirds the disc diameter—quite good! In addition, his unit was not sensitive to weather changes.

A different modification was tested by the inventor Lemstrom in his work on electrification of crops (U.S. patents 634,467 October 1899 and 720,711 February 1903). He made his influence machines using counter-rotating drums, instead of discs, for compactness and improved current output. Being compact, Lemstrom's generator could be enclosed in a wood cabinet with chemically dried air. His generator could be run for long periods without maintenance and was not sensitive to weather changes.

In the last modification, I consider a Mr. Schaffers who put forth his theory of how Wimshurst machines work in July 1895 in Vol. 35 of *The Electrician*. His main innovation was to alter the shape and placement of the collector combs.

Even though Wimshurst had experimented with different comb designs, the traditional shape was a simple *U* placed in the horizontal diameter position as seen in most photos. Schaffers skewed the comb arms by about 60 degrees from each other after finding the best places on the two discs to pick up charges. Current output was improved. Most importantly Schaffers' discovery did illustrate the considerable lack of agreement on generator operation theories. Of course, a good theory would lead to improved efficiency, greater voltage, and current output.

4
The author's generator

THE FOLLOWING IS A DETAILED DESCRIPTION OF MY 14-INCH INFLUENCE machine (Fig. 4-1), which is modified in several ways from James Wimshurst's original design. Please review Fig. 3-1, to familiarize yourself with the parts of the generator that are described in the construction notes.

The design concept is very flexible. The discs of the machine can vary from 12 to 14 inches in diameter so experimenters can use inexpensive 33 RPM records for test purposes. These record discs can be used as a substrate for coatings and varnishes. The collector supports are made to permit repositioning of the charge collectors for different disc diameters.

As designed, my machine with Leyden jars can produce sparks up to 10 1/2 inches overall length, using 14-inch diameter acrylic discs. This is three-fourths the disc diameter, which is quite good! Both the Holtz and the Wommelsdorf generators had good sparking distances, normally two-thirds the disc diameter. The 10 1/2 inches is close to the design limit for machines in open air because the positive and negative collectors are 11 inches apart at the nearest point.

Special Note: Those who want to copy my results should duplicate both the dimensions and the materials I specify. Any departure can cause high-voltage leakage or reduced current output and spoil your enthusiasm. Once you have made a good unit with comparable results, then you can begin to experiment on the infinite changes possible in design.

4-1 A modified sectorless Wimshurst generator with 14-inch acrylic discs.

The overall dimensions of the generator's wood stand were sized to fit the drive motor, which is a 12-volt dc dual-shaft unit. Mine was purchased from Jerryco (see parts suppliers in Appendix A). I chose 12 volts dc because such motors are quiet, powerful, and small, and in a pinch, the machine can be battery powered. So, if you live in a heavily populated area and have complaints about static or noise from the folks next door, you can drive to remote areas and power your generator through the vehicle's cigarette lighter outlet.

In operation, the motor starts by drawing 10 amps, which drops to 7 amps when the generator is charged and running at full speed.

My motor has a body diameter of 3 inches and a shaft measuring 5/16 inch in diameter by 8 inches in overall length. The disc size, 14 inches in diameter, is the largest that should be used with this size motor; otherwise losses from air drag become excessive.

If you choose another dual-shaft motor with longer shafts, have the excess shaft length cut off at your local machine shop. Be sure to buy a motor having the same power capacity: 12 to 16 volts dc and 10 amps current when loaded. The wooden parts of the machine should be made from well-seasoned hardwood, such as maple, walnut, mahogany, or oak.

The stand

The stand consists of two side rails measuring 3/4 inch × 1 3/4 inch × 14 inches long, and two end pieces of 3/4 × 1 3/4 × 10 inches long. To fancy up the appearance of the stand, I chose a 5/16-inch-radius "corner rounding" router bit (Fig. 4-2). I don't have a router, so I cut the moldings using the highest speed on my drill press, taking several light cuts. The rails and end pieces are doweled and glued together (no nails). For the feet, I used 2-inch-diameter maple balls (Fig. 4-3), each one joined to the base with a 3/8-inch-diameter dowel pin. You can stick felt pads on the bottom to avoid marring table tops.

4-2 The generator stand with router bit shown for molding cut.

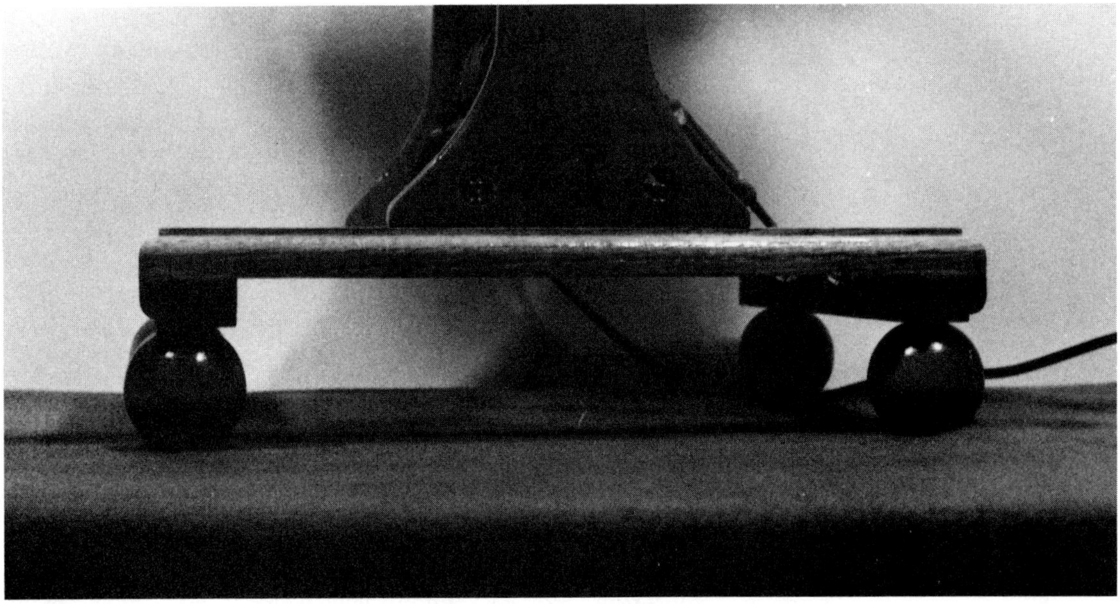

4-3 The generator stand (side view).

The two *stanchions* or upright support arms, are each cut from 1/2-×-6-×-15-inch stock and are attached to the stand with mortise-and-tenon joints or with flathead machine screws and knurled nuts. Attach these arms so they can be easily removed to facilitate changing the machine discs.

Support shaft for discs

This is a 5/16-inch-diameter-×-13-inch-long "drill rod" (ground and sized for use with bushings). If in doubt, take along a 5/16-inch bronze bushing to slide on for checking the fit. You need four 5/16-inch shaft collars to lock onto the shaft with setscrews and allow for positioning the shaft on the support arms. The shaft does not turn; only the bosses and discs do so.

Bosses for the discs

The *bosses* are the round wood projections that attach to the discs on one end and have a *V* groove cut in the other end for the drive belt (Fig. 4-4). They must be well seasoned to avoid cracking, and of a fine-grained wood—hard maple or walnut, for example—to take a high polish. This is one of two generator parts I had made by my machinist, Mr. Riggs, on his metal lathe.

4-4 Boss for supporting disc.

The rough stock dimensions are 4 inches square by $3^1/_4$ inches long for each boss (Fig. 4-5). The first step is to round off the corners to make a cylinder, chuck it in the lathe, and drill through with a $^3/_8$-inch drill bit. Next, moisten the hole with water and quickly press the stock onto a $^3/_8$-inch drill rod, approximately 5 inches long. Chuck the drill rod—the collet chuck or between centers being the most accurate. The moist wood will swell for a tight fit. Turn the outer shape with several light cuts, face-off the ends, and sand the boss to a smooth finish. When the hole is dry, gently tap out the $^3/_8$-inch rod (see Fig. 4-5).

The importance of accurately machining the bosses is a point worth stressing. When mounted with the bronze bushings on the $^5/_{16}$-inch drill rod, the bosses must spin true. The boss faces, on which the discs are mounted, must be carefully machined on the lathe so that the discs spin true. This largely determines a steady reliable generator output.

When the bosses are finished, press into each end of each boss a $^5/_{16}$-inch inside diameter (i.d.) by $^3/_8$-inch outside diameter (o.d.) by $^3/_4$-inch-long bronze "oilite" bushing (available from bearing suppliers). *Oilite* means that the bushing is permanently lubricated with oil. To ensure bushing alignment, you can use a $^5/_{16}$ inch diameter by 4-inch-long drill rod as a guide. I used a large vise to press them in.

The bushings should fit tightly in the $^3/_8$-inch hole, and each boss must spin easily on the support shaft. Drill four holes for attaching the disc to the boss face. I prefer flathead drywall screws with a $^7/_{50}$-inch

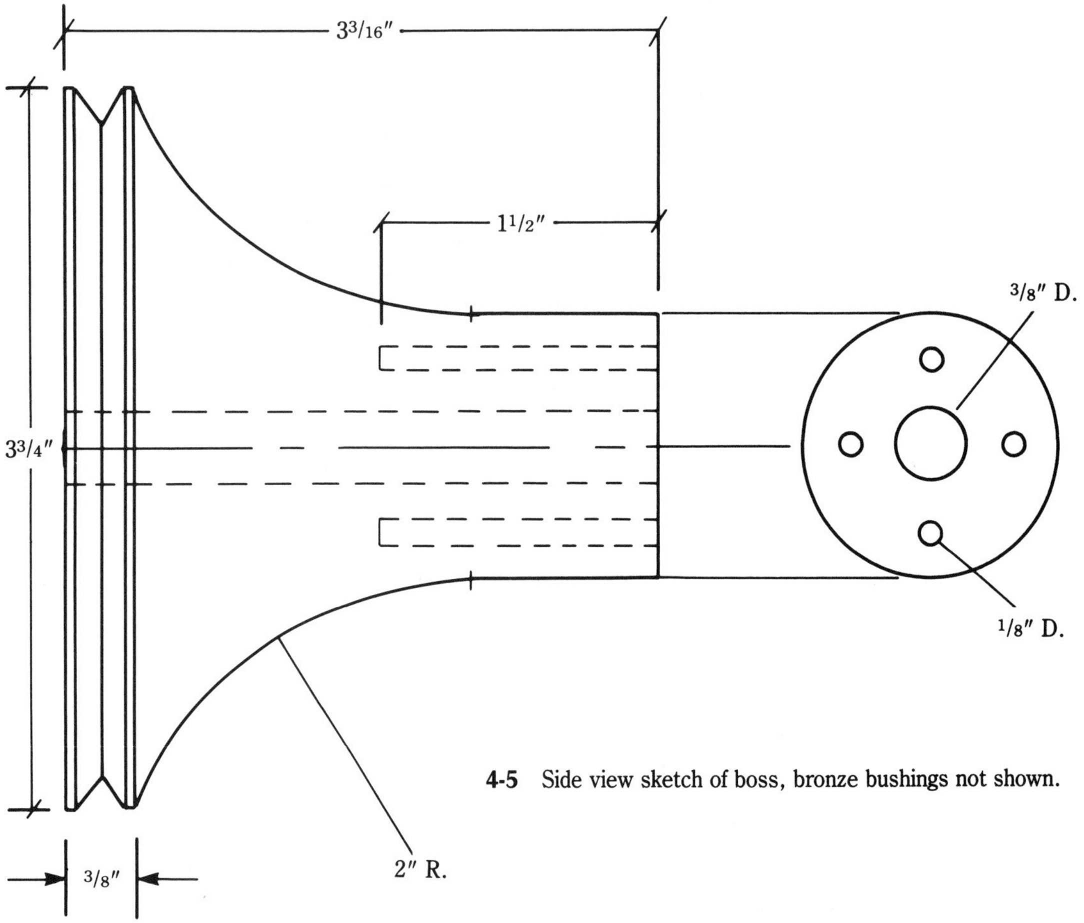

4-5 Side view sketch of boss, bronze bushings not shown.

thread diameter and 1 1/2 inches long because the threads are sharp and deep, and will not strip out easily. The holes in the boss should be slightly smaller in diameter—1/8 inch in diameter is a good size, but not smaller; otherwise, you might crack and ruin the boss. Use a scrap piece of the same wood and test a screw to full depth, if in doubt. The fit should not be tight, just snug, since the discs are light in weight. Use beeswax or bar soap on the screw threads for lubricant, if needed.

Give the bosses three light coats of high-gloss varnish to finish them.

Drive belts and pulleys

The drivers are 3/16-inch-diameter leather-treadle sewing-machine belts, and are available at fabric shops. Cut the drivers to length, and join them by drilling a 1/16-inch hole through the cut end and bending the steel clip provided to make a simple butt joint (see Fig. 4-6). Leave some slack in your two belts and remember that one belt is crossed so the two discs will spin in opposite directions. Control belt tension, as shown in Fig. 4-7, with the two motor turnbuckles.

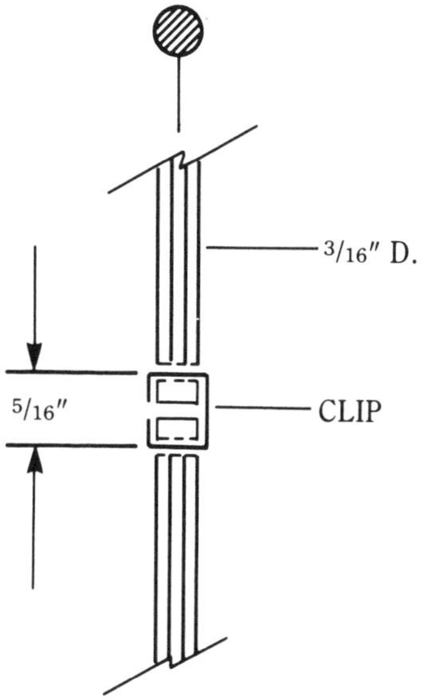

4-6 Drive-belt splice.

4-7 Drive belts and steel pulleys.

The two drive pulleys on the motor (Fig. 4-7) were machined from turned, ground, and polished steel, SAE #1045. This steel is more resistant to rust than cold-rolled steel. Each pulley is 3/8 inch thick by 2 inches in diameter (Fig. 4-8). The pulleys are drilled and reamed for the 5/16-inch motor shaft and are locked in place with a single #10 – 32 setscrew (see Fig. 4-8). The drive pulleys should not be made larger in size for this motor.

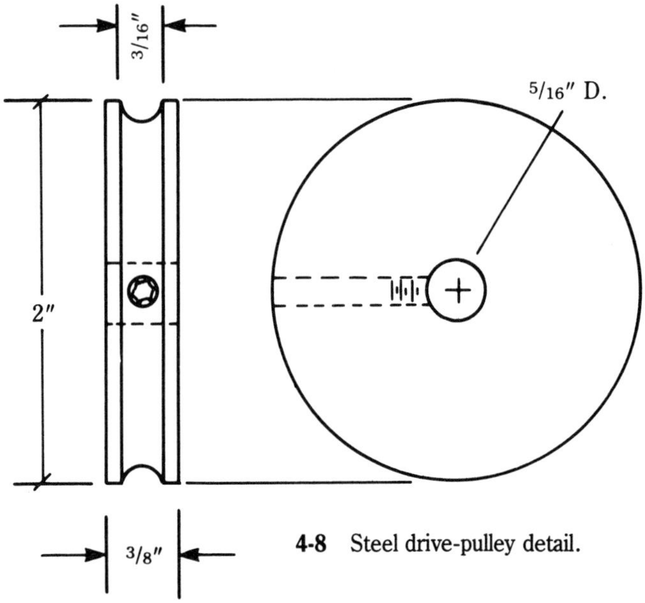

4-8 Steel drive-pulley detail.

Belt tension and motor mount

Attach the two hook and eye turnbuckles, with a maximum open length of 5 1/2 inches, directly to the motor casing bolt. Anchor them at the other end with two eye screws, 1 1/4 inches long, sunk into the end of the generator stand.

Mount the motor (Fig. 4-9) so that it slides from side-to-side on a horizontal brass rod measuring 3/16 inch diameter by 8 3/4 inches long; a brazing rod from the welder's shop will do. Fasten two flat brass washers with a 3/4 inch o.d. to the motor casing bolts, and let the horizontal rod into each upright support arm with a 3/16-inch blind hole.

Charge collector support system

The charge collector support system is mainly acrylic plastic (Fig. 4-10). The two horizontal beams are each 1/2 × 1 1/4 × 16 inches (Fig. 4-11).

4-9 Dual shaft motor and mounting.

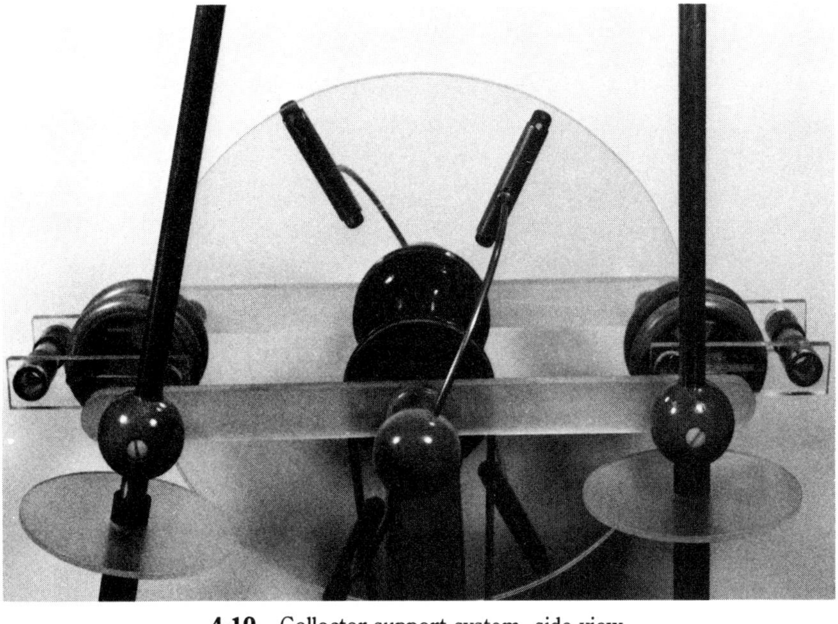

4-10 Collector support system, side view.

Let into each end are acrylic cranks, which hold the collectors. One arm of each crank has a hole, through which passes a $5/16$- × -$3\frac{1}{2}$-inch steel lag bolt. Use the bolt to attach a 2-inch-diameter wood ball. These balls hold the two discharging rods. Running diametrically through each

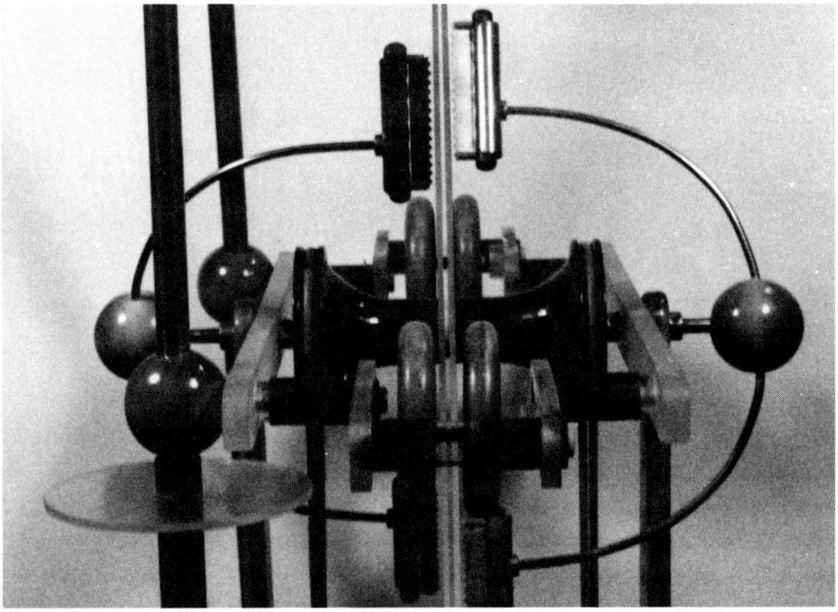

4-11 Collector support system, end view.

ball is a $^5/_8$-inch diameter hole. Rigid copper tubing, $^1/_2$ inch i.d., passes through the ball and supports a discharge terminal. You can slide and lock the terminal rods in any position with a single nylon tightening screw. Do not use metal screws here since they cause leakage (see Figs. 4-12 and 4-13).

Drill all large holes in the plastic parts with wood-boring or "paddle" bits. Always test holes for fit by using scraps of plastic, and use a slow speed to avoid melting problems. Should a hole be too big, simple grind off, equally, both sides of your boring bit until the hole is properly sized. When drilling large holes, always clamp plastic parts to the drill-press table and use a backup piece of wood.

Cement plastic parts for each crank together with acrylic solvent cement, which will flow easily into the joints and set up in several hours. Cement can be applied with needle injectors or with a toothpick; be sure the crank parts are square with each other to prevent binding. Figures 4-14 through 4-16 show details of acrylic crank assembly.

Charge collector combs

I returned to Karl Winter's simple design, which reduces leakage and produces the longest sparks. The dual rings of each comb are made from the rim of spoked wooden toy wheels, $3^1/_2$ inches o.d. by $^5/_8$ inch thick. These wheels can be bought from the toy suppliers listed in Appendix A.

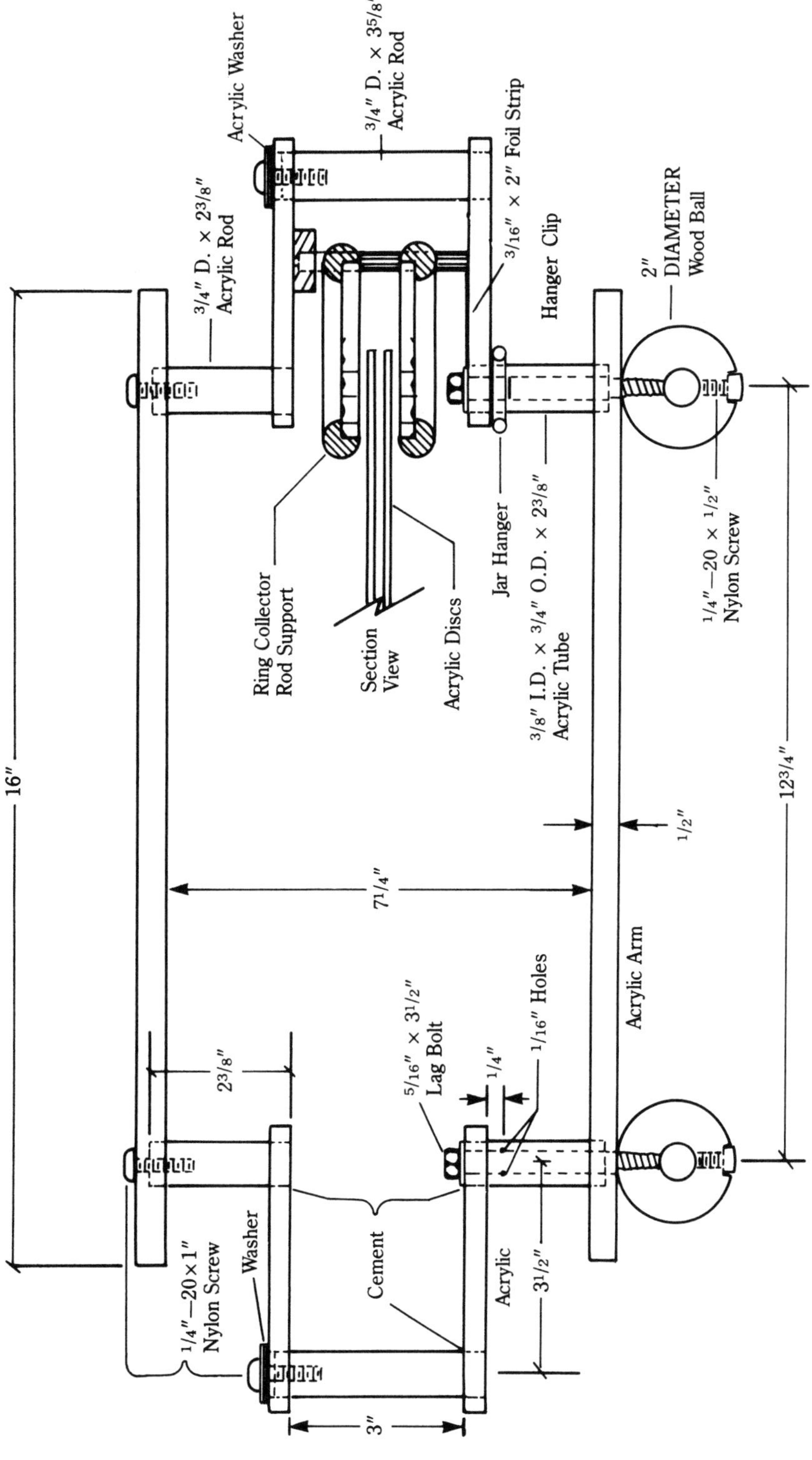

4-12 Collector support system, top view, construction drawing.

4-13 A 2-inch wood ball, foreground, showing the nylon tightening screw.

4-14 Acrylic crank and collector combs, assembled.

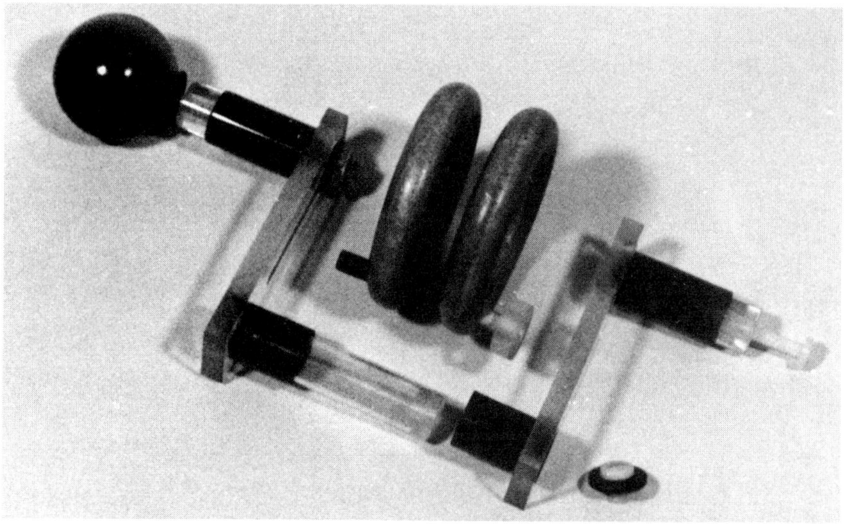

4-15 Acrylic crank and collector combs, disassembled.

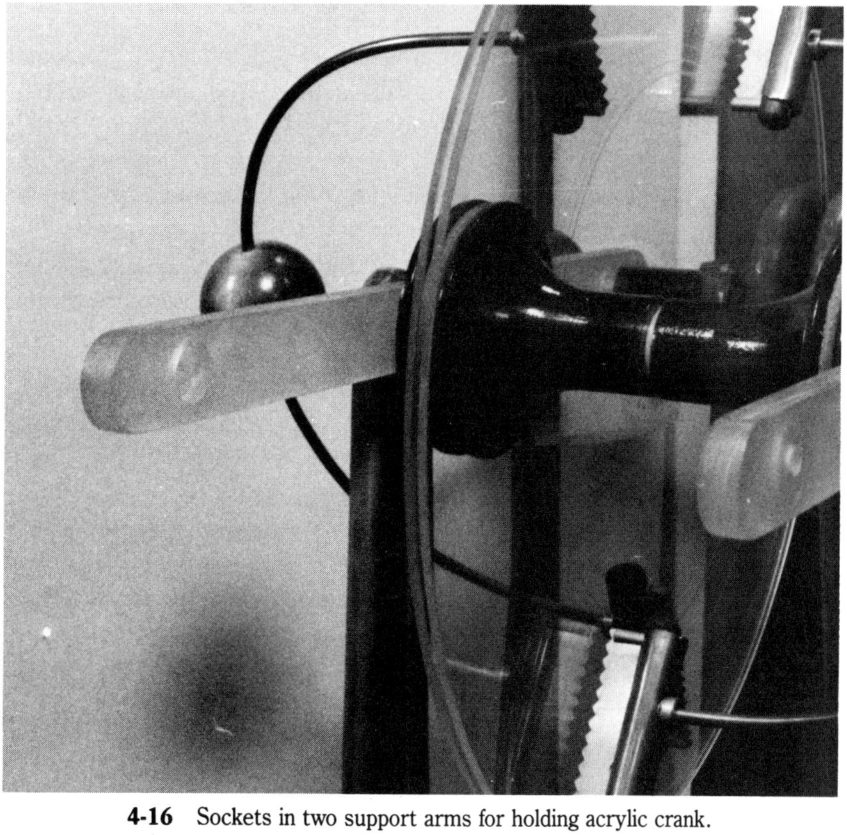

4-16 Sockets in two support arms for holding acrylic crank.

4-17 Dual collector combs made from toy wheels.

Construction of Winter's collecting wood rings begins by knocking out the glued-in spokes and hubs of the toy wheel (Fig. 4-17). Using a 1-inch-diameter hole saw, remove excess glue from the wheel's recess and sand down to the wood.

Set your drill-press table at approximately 4 degrees off horizontal (see Fig. 4-19). Center-punch the highest point on rim and drill through with a 1/4-inch drill bit. Test the hole for perpendicular positioning with a 1/4-inch dowel rod. If the inserted rod is perpendicular to the plane of the wheel rim, finish drilling with a 5/16-inch drill bit at the 4-degree angle. Select a straight section of 5/16-inch dowel rod and cut off two pieces 2 3/4 inches long. You can square the ends of these dowel pieces by placing each in the drill-press chuck and bringing it down against a piece of fine sandpaper on the press table. Coat each dowel rod with graphite for a distance of 2 1/8 inches from one end only.

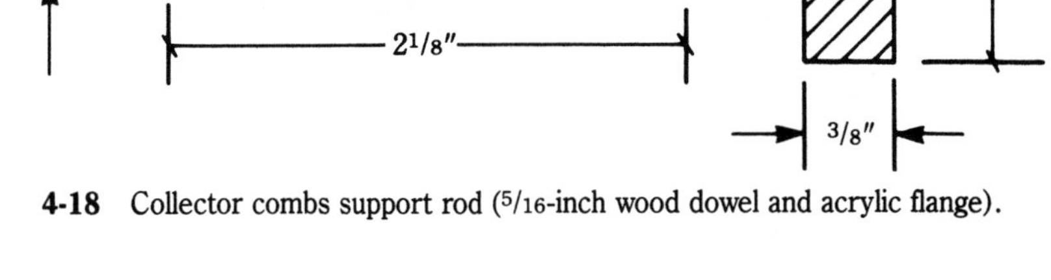

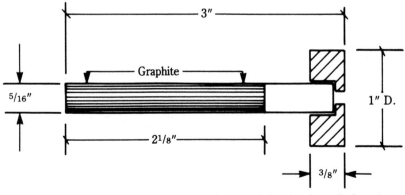

4-18 Collector combs support rod (5/16-inch wood dowel and acrylic flange).

Make a single flange of acrylic for each dowel, cut from the plate with the hole saw. Each flange has a 1-inch diameter and a 5/16-inch-diameter hole 1/4 inch deep, into which the uncoated end of the dowel rod is slipped (see Fig. 4-18). The two pairs of wood rims should slide over their particular dowel rod and hold their positions without binding. The total length of dowel and flange should be 3 inches.

Finish each of the four wheel rims by fine sanding, staining if desired, and applying two coats of glossy polyurethane varnish. Only the recessed area should be left unvarnished so that glue can later be applied.

Next, cut brass or aluminum discs from shim stock, approximately 1/200 inch thick and 2 3/4 inches in diameter, and spot-glue these discs inside the recess in each wood rim (see Fig. 4-19). On the inside of each disc, apply epoxy glue to 22 or 23 brass thumbtacks (3/8-inch head diameter and 3/8-inch tack length). The tacks will cover about half the metal disc's diameter and will give you an area of sharp points for picking up charges from the electrified discs. The ring-shaped combs help prevent charge leakage. Each pair of combs should be set as shown in Fig. 4-14 so that its tack points are greater than 1/16 inch, but closer than 1/8 inch away from the disc's surface. Both rings must be parallel to each other and to the acrylic discs.

Notice that only half the comb's area covers the discs. Trying to cover more area will increase current output, but will also lower the voltage because of excess leakage. This leakage is easily seen in the dark. Clamp the comb sets and dowel supports between the acrylic arms of each crank (see Fig. 4-15). The graphite-coated portion of each dowel

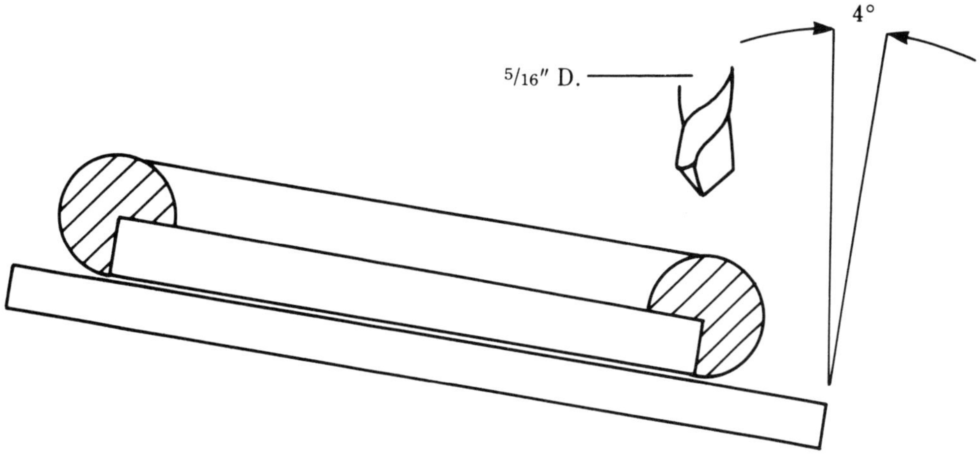

4-19 Drill-press setup for wheel rims.

rod will contact an adhesive foil strip ($3/16 \times 2$ inches) that runs inside the crank arm to the terminal lag bolt (see Fig. 4-12). A good electrical circuit, from combs across the crank arm through the lag bolt to the discharger rods, is now established.

Discharge terminals, supports, and rods

The terminal rods, as explained earlier, pass through the 5/8-inch holes in the two wood balls and are locked in place with 1/4-inch 20 nylon screws. Make the two rods from a 1/2- × -12-inch rigid copper tube. Take great care with the metallic parts of the terminal system to remove all burrs, round-off all sharp corners, and finally, buff these items to a high polish (see Fig. 4-20). Otherwise, leakage will result, causing shorter spark lengths.

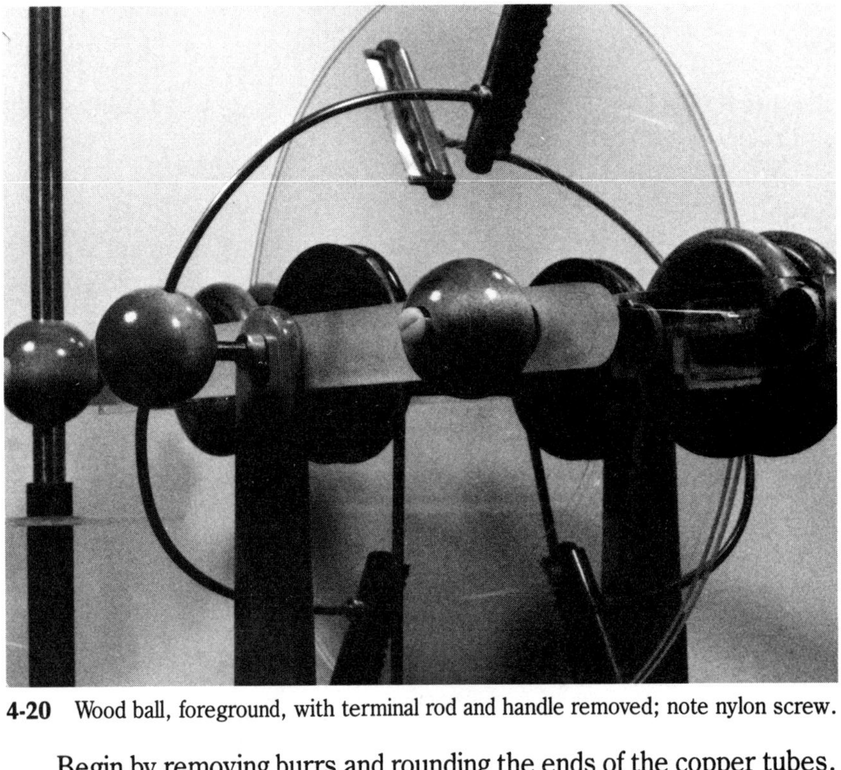

4-20 Wood ball, foreground, with terminal rod and handle removed; note nylon screw.

Begin by removing burrs and rounding the ends of the copper tubes. Into one end of each tube, solder a 1/4-inch 20 stove bolt so that the threads are exposed 3/8 inch (see Fig. 4-21). Fill in the gap with solder until the solder is level with the end of the tube. Use a fine file to level the puddle. Clean off the soldering flux and polish these two tubes.

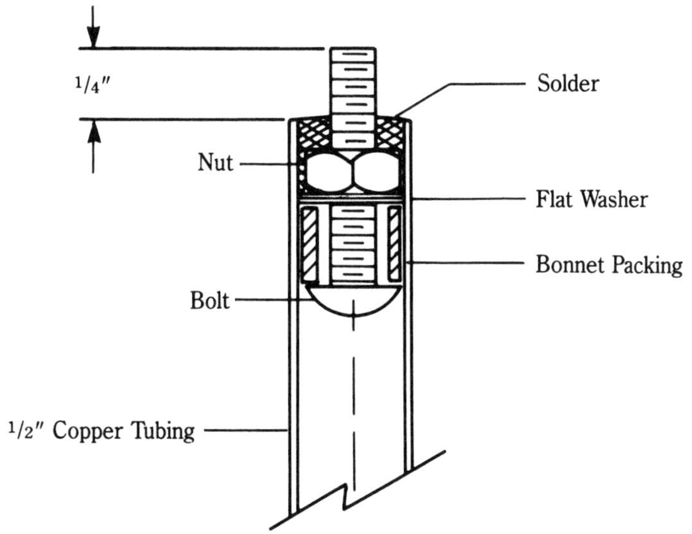

4-21 Stud bolt installation for mounting terminals (*Bonnet Packing* is a plumbing hardware item).

Make the insulating handles from acrylic tubes 5/8 inch i.d. by 7/8 inch o.d. by 12 inches long. Don't skimp on handle length! Onto the top end of each handle glue spark shields as an additional safety factor. Cut each shield from 1/4-inch plate acrylic with a circle cutter on the drill press; each shield is 5 inches in diameter. Clamp each disc down and bore a 7/8-inch hole through the center with a wood boring bit. This shield must fit tightly on the handle, 3/4 inch from the top end. When the shield is smooth and installed squarely in place, run acrylic solvent cement around the joint from top to bottom (see Fig. 4-22).

Slip the shielded end of the handles over the 1/2-inch copper tubing for a length of 1 inch. Enlarge the acrylic tube's hole by sanding if the fit is too tight. The fit should be snug, but the handles should be removable.

The high-voltage terminal ends are normally ball-shaped, but oblates, egg shapes, and rings can be used as well. In Fig. 4-23, the single ball is 3 inches in diameter; the double ball shown includes a 3- to 3 1/2-inch-diameter ball on which is screwed a 1 1/8- to 1 1/4-inch-diameter chrome steel ball bearing. In most cases, nonsymmetrically sized terminals give longer sparks than two round terminals of the same size. The size combination given will produce 10 1/2-inch-long sparks, with the 14-inch-diameter acrylic discs.

Use your imagination when shopping for terminals—doorknobs, drawer pulls, brass bed post caps, fireplace andirons, and float balls have been used for terminals. A good supplier for float balls is Arthur Harris and Company, which carries several shapes and sizes. I prefer #304 stainless steel float balls, which have smooth seams, polish well, and don't tarnish. High polish is important here.

An advantage to using floats is that they have threaded holes. Always request an internal connection with 1/4-inch 20 size thread—this internal thread reduces leakage. When making ball terminals smaller than 2 inches in diameter, I use chrome steel ball bearings.

A threaded connector for ball bearings can be made from ***tee nuts*** or *tee fasteners*, which are available in the nuts and bolts section of your hardware store (see Fig. 4-24).

Place the desired ball bearing against the flange of the tee nut, as shown, and use a 3/4-inch 10 hex nut as a backup. When these three are pressed together in a vise, the flange is bent to the curvature of the ball.

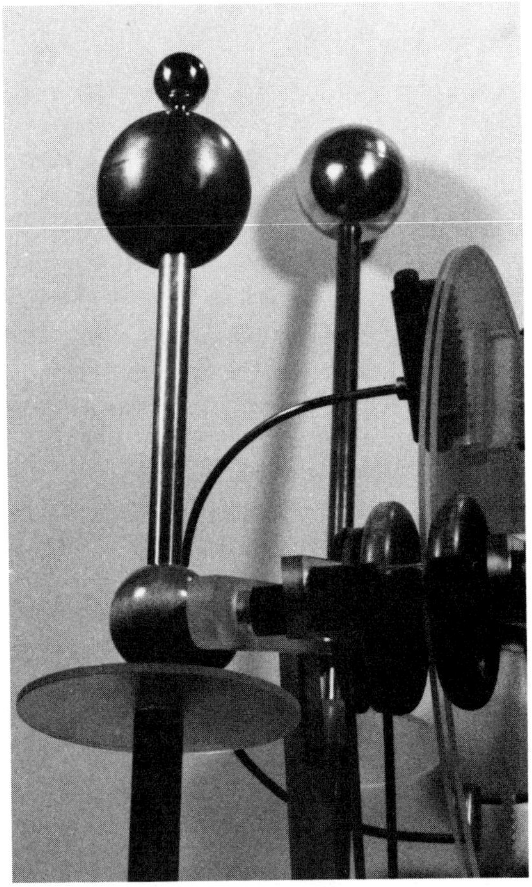

4-22 Terminal support rod, insulating handle and spark shield, installed.

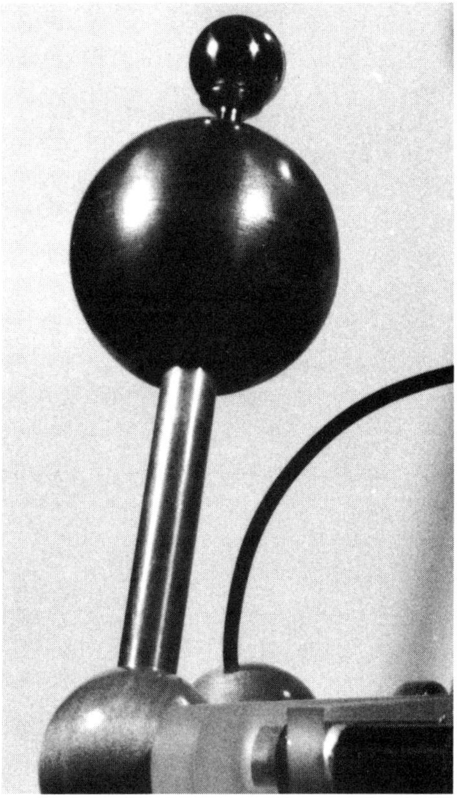

4-23 High-voltage terminals. Note that edge on hole is turned in, to reduce leakage.

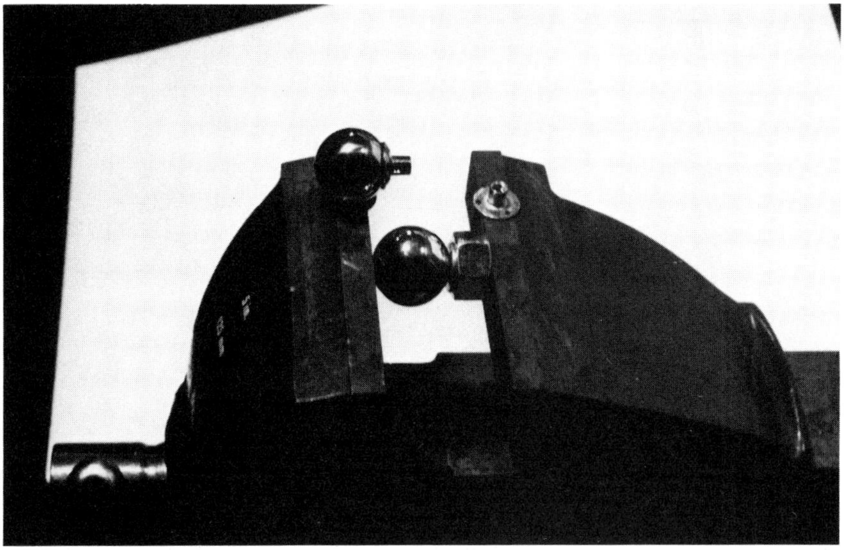

4-24 Forming *tee nuts* in the vice for soldering to ball-bearing terminal.

Clean both parts in alcohol, and tin the flange and ball separately with 40/60 acid-core solder. Turn your modified tee-nut flange up in the vise, place the ball on top, and gently heat only the ball with the propane torch. Do not overheat; the ball should turn slightly purple and will nestle into the flange cup. Cool with dripping water until cold, then clean off the flux. Remove any excess solder. Mount the ball terminals on a 1/4-inch threaded rod and buff them to a high polish; they are now ready for use. I modify the 3 1/2-inch float ball for attaching the smaller ball (see Fig. 4-25).

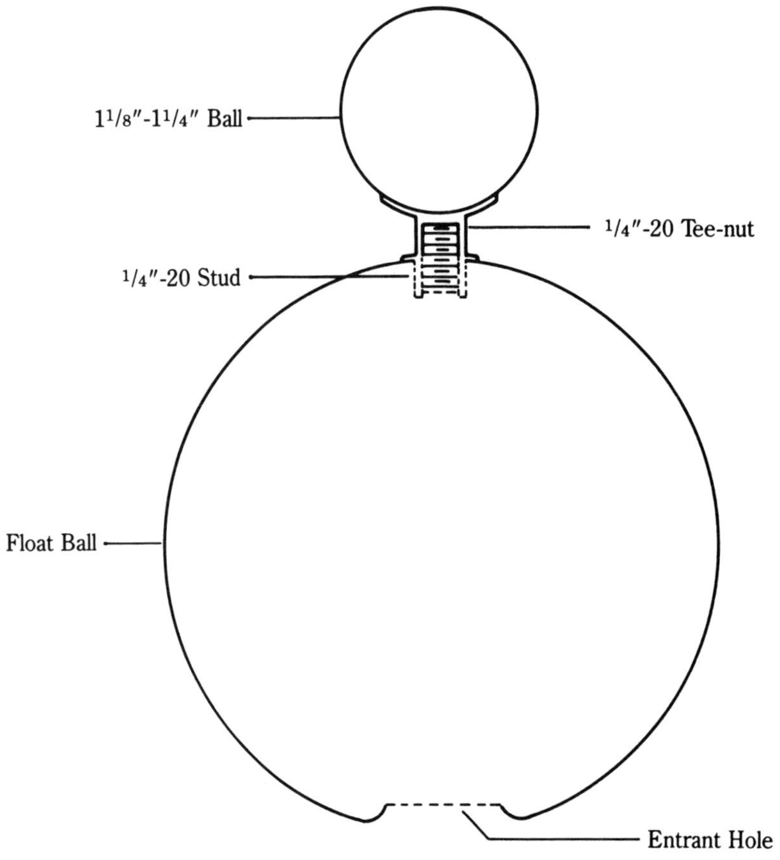

4-25 Positive double-ball terminal.

Notice the turned-in edge of the entrant hole. Cut a 3/4-inch hole in the float opposite to the internal connection. Sand the edge smoothly, removing all burrs. Next, get a small ball-peen hammer. While holding the float in your lap to cushion it, tap the edge the hole is in while turning the float. The edge must be turned inward as shown in the figure. Bore through the internal connection with the 1/4-inch 20 tap drill and tap all the way through it. Place a short threaded stud in the tee nut to complete the double-ball unit, which is assembled as shown in Fig. 4-26. All parts must be highly polished and without burrs. Figure 4-26 shows the negative terminal installed.

Neutralizer rods and blades

Figure 4-27 shows a 2-inch-diameter maple ball with a $^{3}/_{16}$-inch hole drilled through diametrically, and from one side at the 90-degree position, a $^{5}/_{16}$-inch hole drilled to the center (see Fig. 4-28). The $^{5}/_{16}$-inch hole allows you to mount the neutralizer unit on the $^{5}/_{16}$-inch main shaft. Finely sand, stain, and give the two maple balls two coats of glossy varnish. Procure a $^{3}/_{16}$-inch solid (bare) brass brazing rod at the welder's supply shop and bend a section in, to a quarter circle. Approximately $^{1}/_{2}$ inch of each rod will stick inside the wood ball, and the overall length of each brass rod will be 8 inches. The outer ends must be parallel to each other and $9^{1}/_{2}$ inches apart (see Fig. 4-29). Onto these ends, attach the adjustable blades.

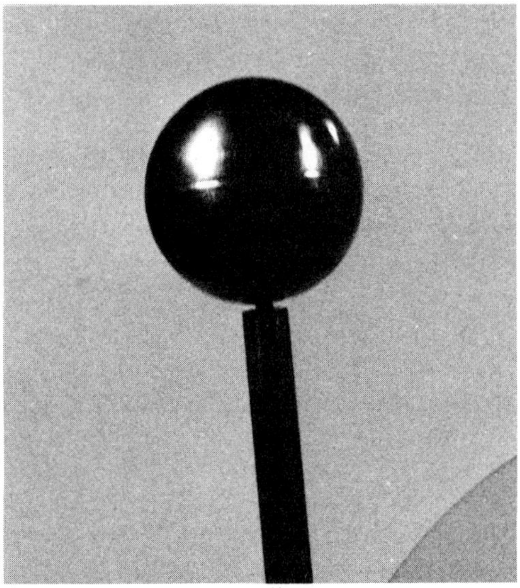

4-26 Negative terminal, screwed out to show stud bolt.

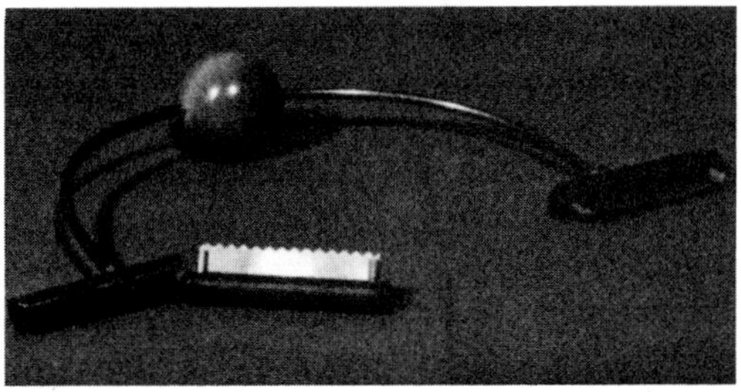

4-27 Neutralizer unit (two required).

The tubular blade holders are 3-inch-long pieces of 1/2-inch rigid copper tubing. Drill a 1/2-inch hole 1 inch from one end. Deburr and solder in a 3/16-inch-wide-slot lengthwise along the tube (see Fig. 4-30) and round the corners of the slot. Deburr and polish each blade holder. Cut a 1/2-inch-diameter by 3 1/2-inch-long wood dowel, and round the ends in the drill press. Then drill two 3/16-inch holes 3 inches apart and insert 3/16-inch dowels. The dowels should protrude 3/8 inch from the surface.

4-28 Wood V-block for center drilling ball (Note centering bushings in foreground).

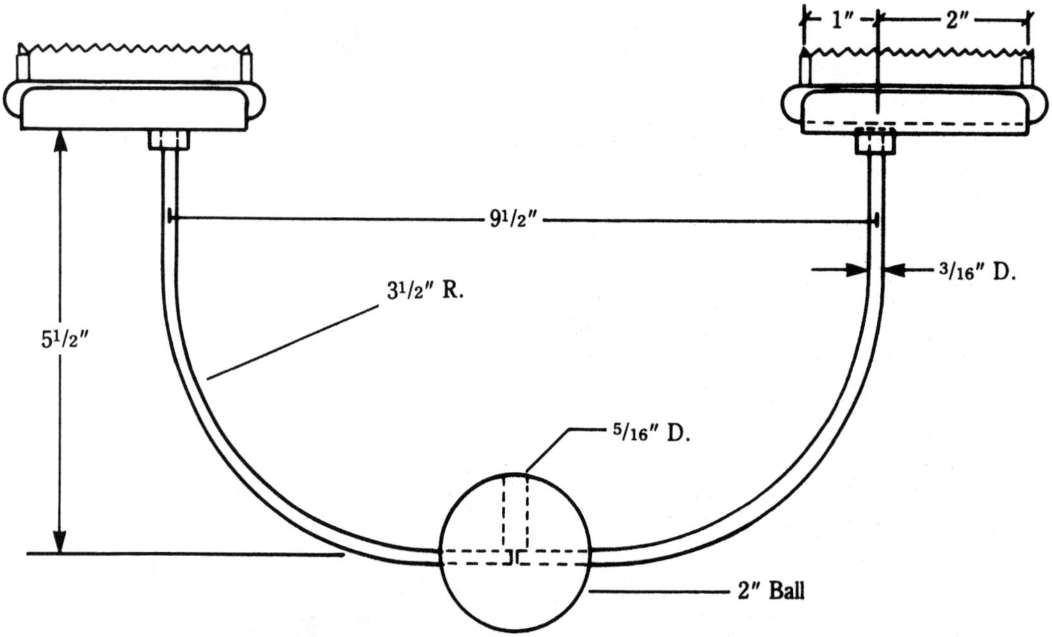

4-29 Neutralizer, construction drawing.

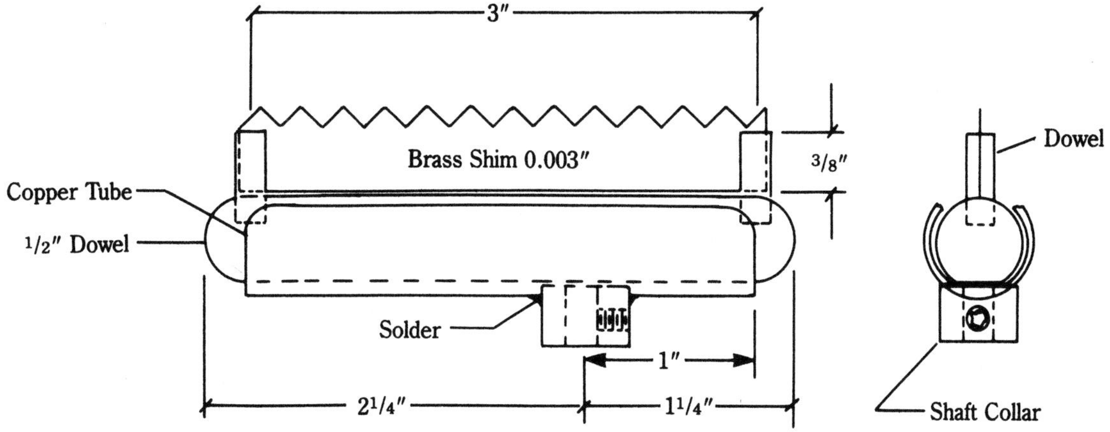

4-30 Neutralizer blade detail.

Split the dowels with a thin knife blade and slip into these split ends a serrated-edge blade made from 0.003- to 0.005-inch thick brass shim that is 1/2 inch wide and 3 1/8 inches long. The points of the serrated edge must be in a straight line. Mine were cut using dressmaker's pinking shears, which should only be used on thin brass shims.

If the blade fits properly, remove it. Next, rub graphite into the dowel parts with a cloth and remove the excess graphite. This procedure will transform the wood into a good semiconductor. Then reinstall the blades and crimp or expand the copper tubing so that your 1/2-inch dowel and blade will slide in, but will hold the position given.

Install the two completed neutralizers. Refer to Figs. 4-27 through 4-30 for the neutralizer placement on the 5/16-inch shaft. The four blades' serrated edges should be approximately 1/16 inch from the disc surface. Bend the brass rods slightly to achieve this positioning. Note that no glue is used for joining the brass rods to the central 2-inch ball. Instead, they can be swung out for individual work on each blade holder.

Now the basic generator is completed to the point that it will charge. Leave the charge-storing unit until later because you need to gain experience until you can work easily around the machine without getting jolted!

12-volt power supply

With a 12-volt dc motor, you can build a simple power supply with a step-down transformer (producing about 10 volts ac at 8 amps) and a bridge rectifier rated for at least 24 volts at 10 amps (Fig. 4-31). Speed control will come from a motor rheostat rated for about 10 amps of capacity. A homemade version is described here. Figure 4-32 shows a simple rheostat consisting of .032-inch (20 guage) by 40-foot-long black steel "stove wire'' wrapped onto a wood skeleton 3 inches in diameter and 6 inches long. About twelve turns are required. Alligator clips permit coil tapping for the desired motor speed. This resistance is only designed for a 12-volt, 10-amp capacity motor. Now that a power supply is complete for the motor drive, the fun can begin!

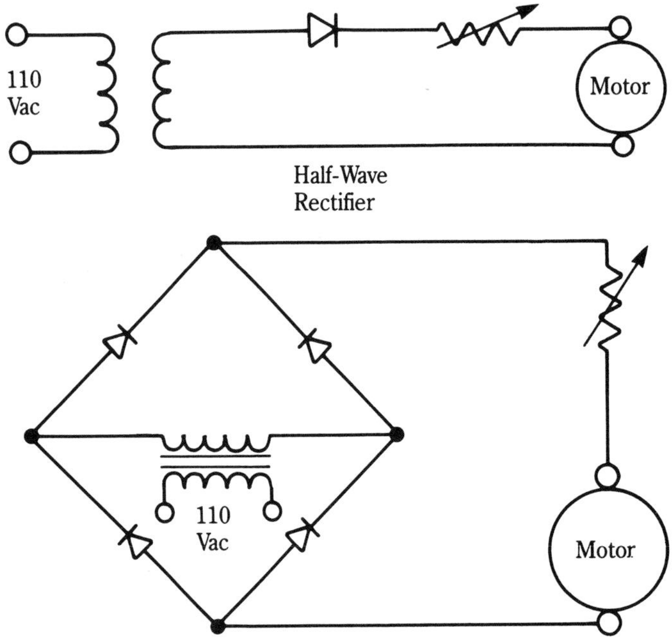

4-31 12 Volt dc power-supply circuits.

4-32 Variable-resistance motor-speed control.

Charging the generator

First, adjust the turnbuckles so that the two motor belts are fairly loose. They will warm up slightly from running; after a while take up the tension to reduce slippage.

In Fig. 4-33, Diana shows how to charge the spinning discs. First, note the white arrow on the disc, which indicates the motion of the front disc—the back disc moves in the opposite direction. You can see how to place the neutralizers: they form an *X*, and one blade from the backside is next to the head of the arrow. The angle between the two topside blades should be about the same as is shown in Figs. 4-34 and 4-35.

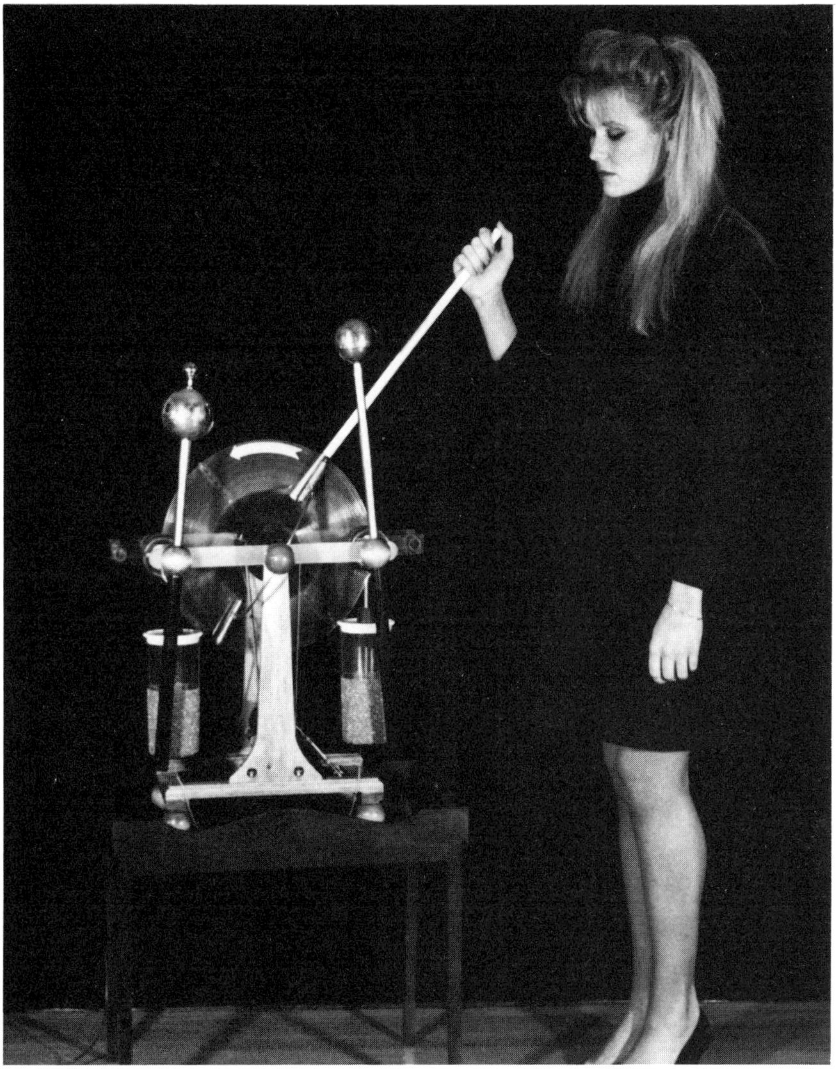

4-33 Charging the generator with PVC pipe.

4-34 Neutralizers at widest angle - 90 degrees.

4-35 Neutralizers at smallest angle - 40 degrees.

The charging tube is a 2-foot length of 1/2-inch PVC pipe. Rub the pipe briskly with a dry clean cloth until it crackles and place it as shown. In Fig. 4-36, seen from the end view, the charged tube is placed parallel to the neutralizer blade and behind both discs, about 1/4 inch apart.

As shown in these photos, the charged PVC tube will impart a negative charge to the single 3-inch ball and a positive charge to the double-ball terminal. Later, when condensers are added, this polarity will produce the longest sparks and highest voltage.

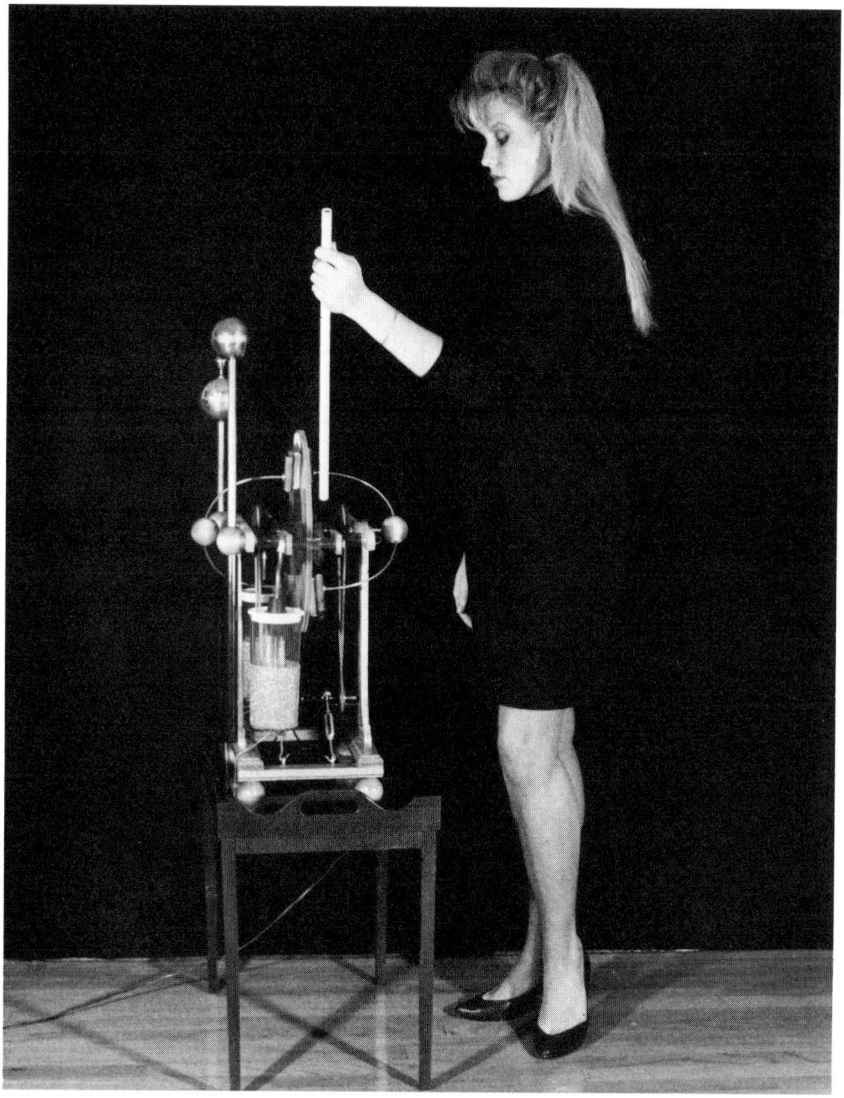

4-36 Charging; place PVC pipe on backside of both discs, next to neutralizer blade.

For safety's sake, try to work on dry and clean wood floors, and stand on a solid black rubber "welcome" mat 18 × 30 inches. Wear shoes with rubber soles, but without nails.

When the machine is charged, its speed will drop slightly and produce a crackling sound; the smell of ozone will fill the air (which you should remove with adequate ventilation). Bringing the terminals together, you should get a quiet brush-shaped discharge about 7 inches long in the darkened room. If the room is very humid (over 80 percent, for example) use a dehumidifier to lower the relative humidity to 50 percent for best results.

When you have experimented for several weeks without being shocked, proceed to build some Leyden jars.

Leyden jar condenser storage

The Leyden jars act like modern electrical capacitors, providing energy storage. The term *condenser* was originally used because at the time electricity was pictured as being like a fluid substance, so the condenser was seen as a storage tank.

I made the jars in Fig. 4-37 using two plastic food storage containers found at kitchenware stores. Dimensions are 3½ inches in diameter and 7½ inches in height with a slight taper. The wall thickness is 1/16 inch. The plastic lids are simply friction-fitted to the jars.

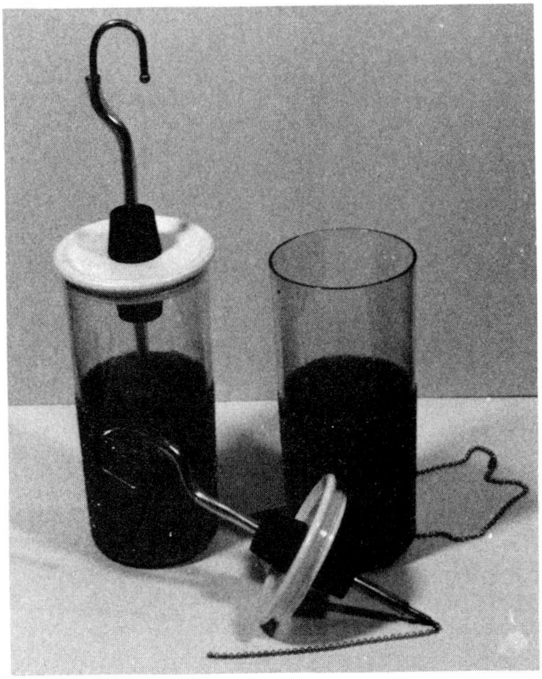

4-37 150,000-Volt Leyden jars made from food-storage containers.

For each jar, bore a 3/8-inch-diameter hole through the center of the lid. Bend a length of 1/8-inch solid copper ground wire into the shape shown in Fig. 4-38. The overall length of the wire will be 11 to 12 inches.

At the end of the hook, solder a 1/4-inch ball bearing or 1/4-inch round fishing line weight. Position a 6 1/2-inch-long section of 3/16-inch i.d. by 5/16-inch o.d. flexible vinyl tubing on the rod as shown. Drill a 1/16-inch hole at the bottom of the rod and solder or clip a 6-inch section of small-link brass or bead chain in place.

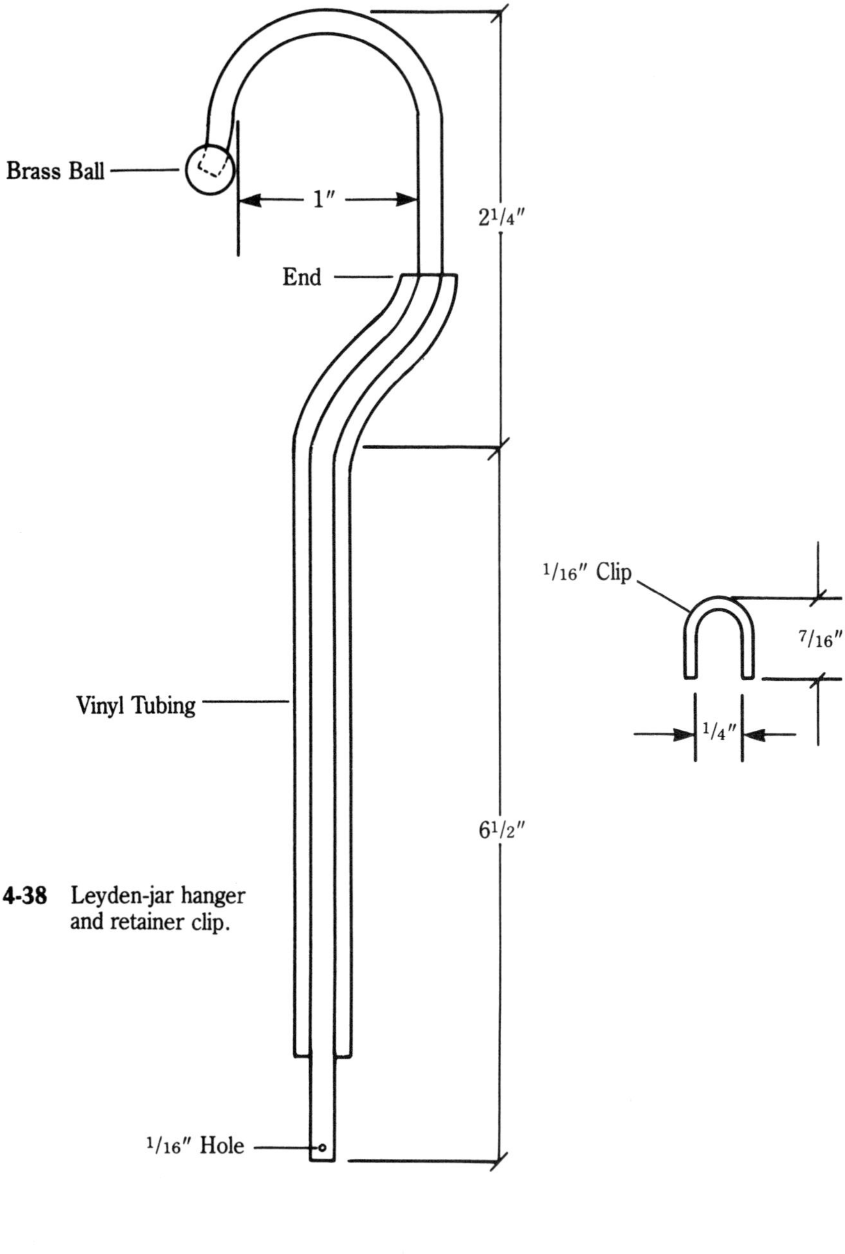

4-38 Leyden-jar hanger and retainer clip.

For each jar, bore a 5/16-inch hole with a tube drill through two #6 rubber stoppers or corks, and slip the stoppers and lid over the rod and tubing as in Fig. 4-37. Do not glue the stoppers and corks (see Fig. 4-39.)

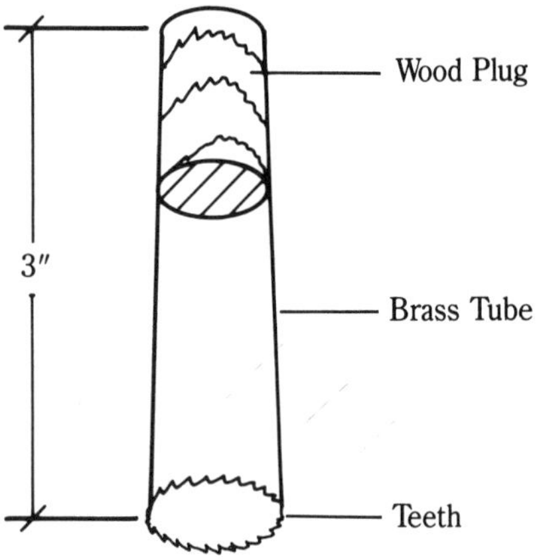

4-39 Homemade tube drill, 5/16 inch o.d., cut bottom teeth with a triangular file.

The coatings for jars of this height should only extend up 4 inches from the bottom inside and out. Clean the jars and place masking tape around the inside circumference above the 4-inch level. The tape will produce a smooth edge for the wood glue. Allow about 1 to 2 ounces of Elmer's wood glue to thicken in the air, until it flows like molasses. Using a brush, evenly coat the bottom and inside of the jar up to your tape line. Remove or spread excess drips of glue and let the jar stand for 8 to 10 minutes, until the coating is tacky. Pour in metallic filings or powder of copper, brass, iron, or lead. Roll the jar by hand until the inside is evenly coated. You can dab any open spots with glue and sprinkle them with the filings. When finished, remove the tape and let the jars stand in open air about one day to cure, then turn the jar upside down and tap it on a table to remove any loose filings. A single coating inside is sufficient.

Now reclean the outside of the jar and place masking tape on the outside at the same 4-inch level. Cut a small disc of adhesive foil to fit the jar's outside bottom. Roughen the outside surface (to be coated) with #220 grit sandpaper and coat the area with wood glue. When the area is tacky, pour filings over it while rolling the jar by hand, with newspaper spread under it. Reuse excess filings until the outside is uniformly covered. Let the jar air dry one day and tap to remove excess metal.

When both jars are completed, tape or glue a length of fine bead or jewelry chain on the bottom. The length of this chain should total 20 inches. When installed, the middle of the chain should make contact with your table to help ground the outer coatings, for safety's sake.

4-40 Leyden-jar hanger and clip installed.

Hang these jars as shown in Fig. 4-40 and keep them in place with a wire clip that passes through the acrylic tube's wall and contacts the 5/16-inch lag bolt.

Because it is often difficult to find food storage containers of the proper shape, I provide an optional schematic for making Leyden jars (see Fig. 4-41). Shipping tubes, made of butyrate clear plastic, with tight-fitting end caps, can be modified for use. A source for these items is listed in appendix A. For each, cut a section of this tubing, 1 1/2 inches o.d. by 1/16 inch thick by 10 inches long. Mask off the bottom inside for a height of 3/4 inch. This portion should be free from glue and metal powder. Next, coat the inside with wood glue, but leave the top 3 1/2 inches of the tube uncoated to prevent sparkover at high voltage.

When this inner coating is tacky, temporarily install the bottom plastic cap and coat the inside, as before, with metal filings. Wait ten minutes, remove the bottom cap, and strip off the masking tape.

Allow this inner coating to air-dry for one hour. Next, clean all pieces of metal powder from the bottom of the tube, run a thick bead of clear epoxy glue around the inside of the lower cap, and install the cap in place. Because the bottom cap seal is the weakest point, electrically speaking, ensure a good seal by holding the tube upright and letting melted paraffin drip down inside until a 3/4-inch-thick plug forms at the base. Set aside for two hours. Coat the outside of the tube as before, starting just above the bottom cap. Leave the top 3 1/2 inches of the tube

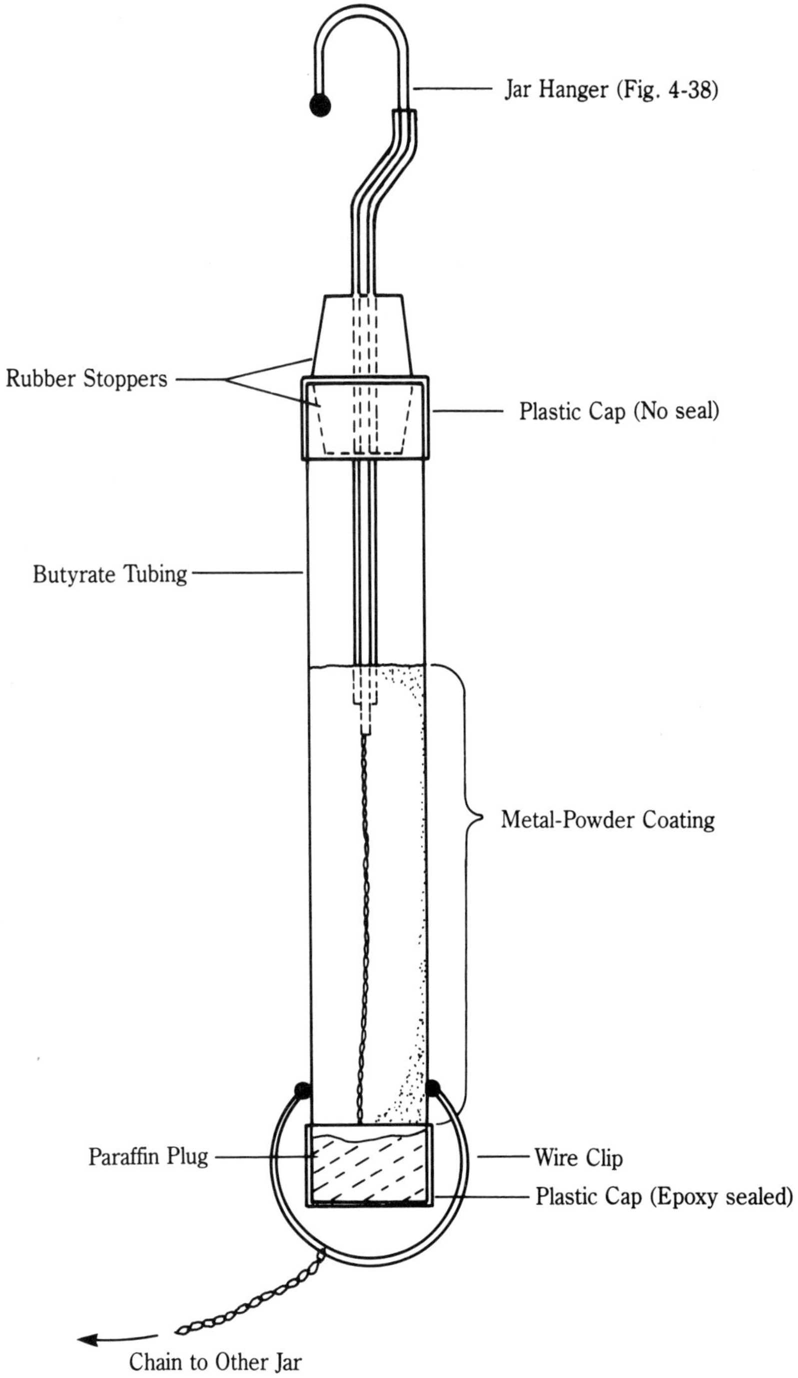

4-41 Optional Leyden jars made from shipping tubes.

4-42 A 10-inch bolt. Discs are 14 inches in diameter Film: ASA 400, f.16, Exposure 4 seconds; Relative Humidity 43 percent.

outside uncoated. Cut a 3/8-inch hole cut in the center of the top cap. Using the same Leyden jar hanger and two rubber stoppers (see Fig. 4-37), assemble the condenser.

Warning: Be sure to keep your hands away from metal parts when charging, and stand on the rubber "welcome" mat. A good practice, recommended by Nikola Tesla, is to use only one hand and keep the other in your pocket. Move only one terminal handle at a time. When finished always touch the terminals together to discharge your stored electricity and leave the terminals together when you are not using the generator. If you have carefully made the machine, as described, you will get 9 1/2- to 10 1/2-inch sparks once every 3 or 4 seconds (see Fig. 4-42).

5
Unusual generator designs

EARLIER I MENTIONED AN ADVANCED INFLUENCE MACHINE THAT WAS MUCH more efficient than the Wimshurst machine: the Wommelsdorf generator (covered by U.S. patents 882,508 in 1908 and 1,071,196 in 1913). So, if this machine was such an improvement, why is it not as well known as the other designs? Mainly because of poor timing. From 1900 onward, the momentum in electrical engineering shifted toward electromagnetic generation because of its practical power output. By the time the Wommelsdorf machine made its debut in *Electrical Review* in 1913, there were few articles on the subject. Also, the plastics "revolution" had not yet occurred; ebonite and bakelite were the best insulators on the market, neither of which are resistant to ozone attack.

This machine is an example of how technical innovations fall into oblivion because of bad timing, unfavorable economic conditions, or the lack of materials to make the device practical. This is why it is always a good idea to review older innovations and reassess them in light of your present situation. There are quite a few "jewels in the rough" in any branch of engineering that are still waiting to be rediscovered.

The article in Fig. 5-1 is intended to supplement the two Wommelsdorf patents. Table 5-1 gives generator performance for several sizes available about the year 1915.

The popularity of this design continued in Germany into the 1930s. Figure 5-2 shows a commercially produced portable unit, probably used by electrotherapists.

*The condenser machine

The condenser machine invented by Dr. H. Wommelsdorf is a new electro-static machine for the direct generation of high-tension continuous current. While its principle was described some years ago in the scientific press, it has only recently been so improved as to become suitable for practical use in X-ray and other work.

The characteristic feature of this machine is the alternating arrangement of the rotary and stationary disc, which is based on the condenser principle. It will be remembered that in an influence machine the rotating disc is only influenced on one side, the generated electricity being drawn off from the other. The rotating discs of the condenser machine, on the other hand, undergo electrostatic induction on both sides, the electricity produced in them being collected from a groove in their extreme periphery by steel wires penetrating therein. It will thus be seen that each disc, in accordance with the theory of condensers, takes up and supplies twice as great an amount of electricity as an influence machine. Actual experience moreover, shows that the output of the new machine, thanks to some additional advantages connected with the condenser principle, is even considerably higher than could be expected. It is mainly due to the close arrangement of the discs that the condenser machine yields an amount of electricity 20 to 50 times as much as a Holtz-Wimshurst multiple-disc influence machine of equal size, the type so far almost exclusively on the market. Another point of considerable importance is that the condenser machine is absolutely independent of atmospheric conditions.

The sectors of the condenser machine are not attached to the surface of the discs, but embedded in their interior, thus augmenting the output, pressure and self-excitation of the machine, and increasing the life of the discs. In fact, the unceasing inflow and outflow of high-tension electricity from the suction combs, brushes and sectors to the insulating discs was bound in a very short time to destroy even the best insulating material. The stationary discs of the machine are likewise entirely encompassed by insulating material.

The condenser machine mainly comprises a substantial frame closed on all sides which carries in its interior the statical fields, that is, the stationary discs. The rotating discs above described, to which is due the influence effect, are sandwiched in between these fields, thus being also comprised in the interior of the frame. This compact arrangement protects all vital parts of the machine not only against dust, but at the same time against radiation losses which would otherwise be an important item.

That this remarkable simple and efficient machine—as it were, a continually charging and discharging condenser—should have had to wait a number of years before being placed on the market, was due to some drawbacks of its original construction and to the fact that no insulating material suitable for the purpose could be found. In fact, the effects of the electricity and ozone produced by the condenser machine are, just because of its extraordinary output, so powerful that rubber, a substance that could not be dispensed with on account of the high voltage of the machine, would become conductive after a very short time.

The adoption of a certain brand of Bakelite with which the caoutchouc discs were coated all round, first made the condenser machine durable and suitable for commercial purposes. This Bakelite layer—insoluble and brilliant as enamel—endows the discs with a remarkably hard surface, and can only be removed by scraping with a knife, in the form of an amber-like yellow powder.

The arrangement of the current collectors at the extreme circumference of the discs has enabled the spark length of the condenser machine to be

* From *The Electrical Review*, August 29th, 1913.

5-1 Wommelsdorf generator. *Journal of the Rontgen Society*, 1914.

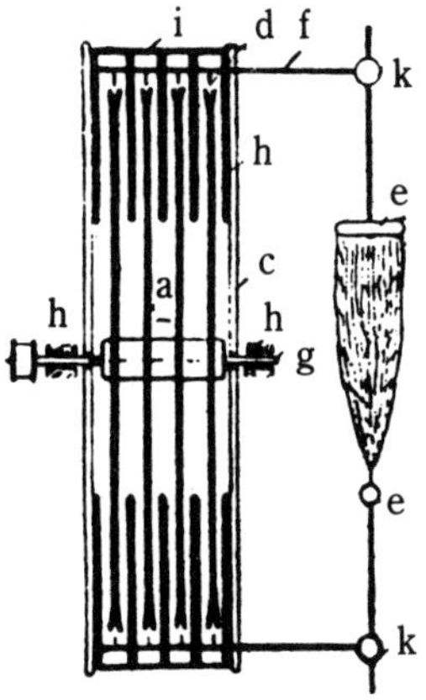

a, Rotating discs; c, end cover; d, collector; i, frame; e, electrodes.

SECTION OF CONDENSER MACHINE.

increased to twice its former value, thus allowing sparks considerably longer than half or even two-thirds of the disc diameter to be produced.

The condenser machine will be found an extremely useful apparatus for X-ray work, affording, as it does, a means of producing high pressure continuous current directly without any conversion, rectification, &c. This allows hand operation to be used in the case of moderate effects, while increasing the life and constancy of X-ray bulbs, the more simple of which can be used without cooling, thus cheapening operation. Wherever there is no suitable electricity supply, *e.g.*, in the case of portable installations for military and other purposes, the condenser machine will afford an ideal means of operating X-ray apparatus. It will also be found useful in cases where only an alternating or three-phase current supply is available, which could not be used without resorting to converters, rectifiers, &c. Apart from hand operation, a small continuous, alternating, or three-phase current motor (of $^1/_5$ H.P.) can be used to drive it. The output of a motordriven condenser machine having a single rotating disc 55 cm. in diameter is as much as $^1/_2$ milliampere. Machines comprising four, eight or ten rotating discs, such as have been constructed, show a correspondingly higher output.

Another field in which the condenser machine is found to prove of the greatest usefulness and to excel all other apparatus so far in use, is the field of electro-therapeutics, viz., the medical applications of electricity. An especially important feature in this connection is that the output of the machine can be controlled at will and reduced to zero, the current intensity being proportional to the number of turns per second. This is an enormous advantage over powerful induction coils, which cannot be used for small outputs.

If, for instance, the electric currents yielded by the machine be applied to the human body, their physiological effects can be altered quite gradually and controlled at will, from a hardly perceptible shiver up to an unbearable intensity. At the same time, the operation of the machine is absolutely safe and harmless, the upper intensity limit being readily adjusted for by choosing a proper size of Leyden jars.

The machine also comprises two special terminals from which alternating currents or rapid oscillations can be collected for high-frequency current work and for performing the methods devised by Oudin, D'Arsonval, Apostoli, &c. The same large condenser machine also lends itself for franklinisation proper, that is, for static electricity treatment, and for experiments with the insulating stool.

THE CONDENSER MACHINE.

The high pressure of the machine is noticed at a considerable distance by a spider-web feeling due to electrically-charged particles. These remote effects can be strikingly demonstrated by lighting Geissler and Tesla tubes from a great distance.

5-1 Continued

The exceptionally high Leyden jars used in connection with the condenser machine are countersunk in its base, and are, like the discs, coated with Bakelite. The polariser at the same time serves as short-circuiter, for rapidly and accurately breaking the flow of sparks.

Though the machines so far constructed are doubtless the starting point of types much larger and more powerful, they are able to comply with the most exacting requirements of X-ray work and electro-therapeutics.

The condenser machine is constructed by Messrs. Berliner Elektros. G.m.b.H., of Schonberg, near Berlin.

5-1 Continued

Table 5-1. Wommelsdorf generator performance.

No. of rotating discs.	Diam. in cm.	Length of sparks in mm.	Current in microamps.	Energy required in H.P.
1	26	170-190	120-140	about 1/16
1	35	210-250	230-280	about 1/8
1	45	260-300	380-480	about 1/6
1	55	300-360	500-600	about 1/6
2	45	260-300	620-750	about 1/6
2	55	300-360	700-850	about 1/4
3	55	300-360	1400-1600	about 1/3
5	55	300-360	2000-2500	about 1/2
7	55	300-360	2800-3400	about 3/4
10	55	300-360	3750-4500	about 1

*(Generator speed for the above was 2300 rpm.)

Modern High-Speed Influence Machines. V. Johnson, 1921

Improved designs

Since the Wommelsdorf machine was invented, three French innovations have improved disc generators.

The first of these innovations by Henri Chaumat; His single-disc generator (46 cm diameter) produced a potential of 300,000 volts. The spark length was 32 cm between terminals 26 cm in diameter. He claimed its output was better than a Holtz or Wimshurst machine's—24 watts at 200,000 volts. The Wimshurst generator delivered 0.7 watts at 70,000 volts. No U.S. patent was received by Chaumat.

The second generator design was developed from Chaumat's prior work and was the invention of Pierre Jolivet in 1953. Jolivet's disc generators used bleeder resistors, by which a portion of the output was used to increase the electric field intensity across the inductors—called *neutralizers* in a Wimshurst machine. No U.S. patent was found for Jolivet's invention, and neither of the previously mentioned articles has been found translated from the original French into English.

5-2 Improved Wommelsdorf. *Elektrotechnische Zeitschrift*, 1929.

The last French innovation appeared concurrently with Jolivet's in the early 1950s. The inventor, Noel Felici, chose to concentrate wholly on cylindrical induction generators and managed to raise the generator efficiency to a remarkable 90 percent, using an environment of pure hydrogen at 15 to 20 atmospheres. A drum generator with a power output ranging from 20 to 3000 watts was feasible with Felici's design. His discoveries so dominated the study of electrostatics that most of the generator patents issued in the United States during the 1950s went to this man. Felici's design is somewhat exotic, and therefore is outside the scope of this book, but it does show the room available for improvement in performance.

Moving in the other direction, toward extreme simplicity, I close this section with two novel approaches. Figure 5-3 shows a device described in 1912.

When I made a modified form of this machine, I substituted two 4-inch-diameter float balls for the discs to reduce leakage. I joined them together to make a dumbbell shape, and used nylon thread in place of linen. A small dc motor and pulley drive the dumbbell at 6,000 to 8,000 rpm.

I used my simple generator to study the influence of static electricity on vortices in air. At full speed, I lower the dumbbell terminal into a wastepaper basket partly filled with Styrofoam packaging "peanuts". Charging does increase the strength of the miniature tornado vortex that forms. My unit also produces sparks 3/4 inch long.

This generator is really like a Van de Graaff generator in its simplest form. An interesting science fair project would be to compare different pulley metals with various kinds of threads—the speed and tension remaining fixed, of course. There should be a variation in potential for each combination of metal and type of thread.

The last unusual generator is one that develops high voltages through the impact of dust particles in a closed system (see Fig. 5-4).

Several impact generators have been patented over the years; some use steam or semiconducting liquids. Voltage and current will vary with air pressure, collector design, particle material, and size. Experimenters wanting to research along this line should provide a way to keep the compressed air dry.

Diatomaceous earth, also used in swimming pool filters, is hazardous when the fine dust is inhaled. Generators that rely on impact electrification come closest to duplicating nature's giant lightning displays. This method of charging is also responsible for fuel transfer and grain elevator explosions; frictional sparks in the presence of finely divided particles in suspension are the main hazard.

Odd static generator

The outfit here illustrated I accidentally discovered will generate static electricity.

Upon a shelf or the edge of a table fasten a small motor that will run at a speed of about 3,00 revolutions per minute. On the shaft of the motor there should be a small grooved brass pulley. Provide a second pulley about 3/4 inch in diameter and a small stove bolt that just fits the hole in it. From thin sheet copper cut two disks exactly alike, and six inches in diameter. Drill a hole in each disk of the same size as the hole in the pulley, and fasten the disks, one on each side of the pulley with the stove bolt as shown. Now make a belt of heavy linen thread. Place the thread over the pulley on the motor and set the disks' pulley upon the loop, thus leaving the disks suspended in the air. Start the motor, give the disks a turn and as soon as they come up to speed, they will act as a gyroscope and keep balanced. Take a piece of copper conductor in the hand and bring the metal near the disks. A 3/4 inch spark can be taken off. An incandescent lamp held near the disks will glow with a weird blue light.—E.H. SAMEN.

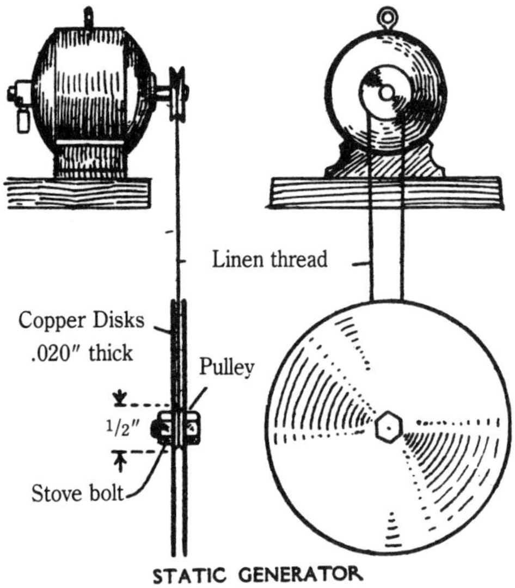

STATIC GENERATOR

5-3 Simple Generator. *Popular Electricity* 1912

Experimental design modifications

Neutralizer units, also named *diametral conductors* or *inductors* depending on theory of operation, are very important. Chaumat, Jolivet, and Felici explored this area extensively. Material, shape, disc area covered, and placement on the disc strongly influence generator efficiency. Some inventors inserted variable resistances between the arms of each neutralizer unit to see how conductivity affected performance. Jolivet, especially, used bleeder resistors or 200,000 to 700,000 ohms to skim off some energy from the collectors to intensify the inductor field strength across the disc.

Experimental disc materials

Some inventors received patents for their work on disc materials. For example, U.S. patent 937,691 by Burton Baker (1909) describes composite discs made from layers of mica and shellac binders. U.S. patent 821,902 by Henry Todd (1906) mentions discs made from fibrous material and treated in a bath of molten sulphur. U.S. patent 1,109,205 by James Dempster (1914) covers discs coated with silica powders to increase surface area and reduce moisture condensation.

Closely related, but not patented, was a description published in 1910. Someone found that if the shellac varnish used by Wimshurst to coat his glass discs is first treated to increase its conductivity, the generator voltages and currents also increase.

The critical process is as follows: Into a clear glass bottle, pour your batch of white shellac varnish. Chop up pure copper wire of small diameter and drop it into the varnish. The varnish, being acidic, attacks the copper. Store this varnish in a warm, dark, dry place for about a week, stirring occasionally. When held up to sunlight, the varnish should show a light green color. At this point, remove all copper to stop the transformation. When allowed to turn dark green, the high copper content makes the conductivity becomes too great.

Coat the glass discs by using Wimshurst's method. Here we see Karl Winter's principle of semiconductor materials showing itself again. The area of disc coatings is the least-explored avenue in this science.

Precision construction techniques

In a more sophisticated direction, Noel Felici (in the 1950s) showed the importance of precision machined rotors and stationary inductors on his electrostatic generator. The traditional influence machines were made with clearances ranging from 0.1 to 0.2 inch between the neutralizers and discs. Felici reduced the working clearance in his cylindrical machines to 0.01 inch, thus permitting a stronger electric field for rotor

A High Voltage Direct Current Generator

By Richard E. Vollrath
University of Southern California, Los Angeles

(Received August 19, 1932)

When powdered materials are blown through metal tubes by means of compressed air considerable quantities of electricity are produced by contact electrification. It was found that 6×10^{-6} coulombs could be produced per gram of diatomaceous earth, a form of silica, blown through a short length of copper tube. A generator of extremely high voltage is proposed, and a small scale model of such a generator is described, by means of which currents of 8×10^{-6} amperes at 260 kilovolts were generated.

THIS work was undertaken to provide numerical data to serve as the basis for the design of a high voltage generator capable of generating a milliampere at voltages above a million.

The Proposed Generator

The discussion to follow will be simplified by a consideration of Fig. 1 which is a diagrammatic representation of the proposed high voltage generator. The small scale model constructed will be described later on.

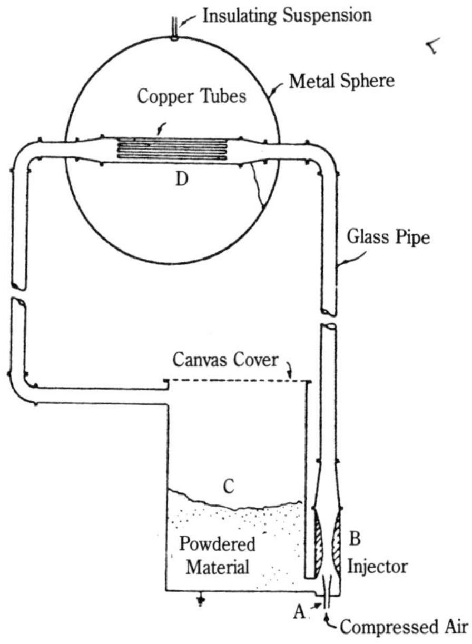

Fig. 1.—Proposed high voltage generator.

A blast of compressed air is blown from a nozzle *A* into a suitably designed injector *B*. A powdered material *C* is sucked into the air stream by the action of the injector and carried along through glass or Bakelite pipe. The air laden with the powdered material passes through a number of metal tubes *D* ar-

5-4 Vollrath's Frictional Dust Generator. *Physical Review*, 1932 ranged in parallel within a large spherical conductor and electrically connected with it. The particles of powder become electrified by contact with the walls of the metal tubes. The charged particles are carried away from the sphere and returned to an earthed reservoir *C*. The potential of the sphere will rise until limited by corona discharge from its surface.

In order that one milliampere can be drawn from the sphere, enough powder must be blown per second to produce a charge of 10^{-3} coulombs or 3×10^6 e.s.u. per second. It was the main purpose of this work to find out if such charges can be obtained from reasonably small quantities of powder. From an engineering standpoint quantities up to about 305 grams per second should be feasible since sand blast machines have been constructed capable of blowing this quantity of sand.

Previous Work

It has been known for a long time that considerable charges are developed when particles of solids are blown over the surfaces of metals and other substances. Most of the work recorded in the literature gives no information as to the quantity of electricity produced in this manner by a given amount of material. However, the following brief resumé of the more pertinent articles showed that the charges obtainable were large enough to warrant further work along these lines.

According to Rudge[1] a few centigrams of flour blown into a large room produced a charged dust cloud whose potential as measured by a radium coated collector, or probe, was 200 volts. He pointed out that this and other results of a similar nature obtained by him accounted for potential gradients of 10,000 volts per meter observed during dust storms, and for lightening flashes occurring during the eruption of ashes from volcanoes.

Petri observed that a steel telegraph wire 5 kilometers long became electrically charged during a violent snowstorm. A continuous stream of sparks several millimeters long could be drawn from the wire; and Petri estimated the electrical power generated to be 1.2 horsepower. The effect has been attributed by Ebert and Hoffmann[2] to contact electrification of the snow blown over the wire by the wind.

A similar observation is recorded by Stäger[3] who exposed a wire 9 meters in length to the driving snow during a snowstorm. There was a distinct corona discharge around the wire and a current of 17 to 20 milliamperes could be drawn from it. In this case the power generated was estimated to be 3 watts. Stäger in the same article gave the charge carried away by hoarfrost blown from a surface of ice. Under particularly favorable circumstances it amounted to 1000 e.s.u. per gram of hoarfrost. He also mentions the appearance of a corona discharge 10 cm long during the production of carbon dioxide snow by rapid evaporation of liquid carbon dioxide escaping from a tank.

[1] W. A. Douglas Rudge, Proc. Roy. Soc. London A90, 256 (1914).
[2] Ebert and Hoffman, Meteor. Zeits. 317 (1900).
[3] A. Stäger, Ann. d. Physik 77, 230 (1925).

5-4 Continued

It is quite likely that the tremendous voltages produced in the Alps, and lately used in attempts to operate large x-ray tubes are generated by the electrification of snow blown over the ice covered peaks.

THEORETICAL LIMITATIONS

In pursuing this work the writer adopted the views of Helmholtz on frictional electricity. According to these, so-called frictional electricity is developed whenever two dissimilar surfaces are brought into contact and then separated. A double layer of charges, whose magnitude is determined by the contact difference of potential between the two surfaces, forms at the surface of contact—one charge residing on one surface and an opposite charge on the other. When the two surfaces are separated the charges of the double layer are torn apart, and a charge remains attached to one surface while the other carries with it a like charge but of opposite sign. A contact of very short duration of two insulators followed by their separation suffices to produce considerable charges, which indicates that the double layer does not penetrate very far into the body of the insulators in contact. The thickness of such double layers has been estimated to be of the order of 10^{-8} to 10^{-7} cm. From this it is evident that in order to produce large charges by contact electrification large surfaces of contact are the main consideration. This immediately suggests that at least one of the two substances brought into contact should be in a finely divided state so as to present a large surface. In this case the charges are produced by blowing the finely divided material over a metal surface; for example the powder is blown through a metal tube. When the particles strike the metal surface and leave it they acquire a charge which they carry with them as they move along with the air stream. An opposite charge remains on the metal which, if insulated, rises in potential as long as the powdered material is blown over it.

Leaving out of consideration corona discharges from the conductor, the ultimate potential which can be reached depends upon the mobility, k, of the charged particles leaving the conductor and the potential gradient, X, at the point where they leave. The charged particles to escape must be impelled by the air stream with a velocity greater than kX. The electrical image force between the particle and the conductor is considered negligible owing to the smallness of the particles under consideration. It can easily be shown that the above requirement imposes no serious limitation upon the potentials attainable, even though the particles should have to overcome the maximum gradient possible in air, about 30,000 volts per cm.

A charged particle of radius r in air will have a maximum mobility when it is carrying the maximum charge q permitted by the limiting gradient at its surface, that is, $q/r^2 = 30{,}000$ or $q = 100\ r^2$ e.s.u. For particles of radius 10^{-4}, $q = 10^{-6}$. Consider 1 cc of material broken into approximately spherical particles 10^{-4} in radius, each charged with the maximum 10^{-6} e.s.u. The total charge carried by all the particles is

$$Q = 10^{-6}/[(4/3)\pi r^3] = — 10^6/4.$$

5-4 Continued

According to this only 12 cc of material would be necessary per second to carry a milliampere. The same calculation for particles of radius 10^{-5} cm gives $10^{-7}/4$ e.s.u. per cc. However, owing to the fact that air in very thin films has a higher breakdown strength, the gradient at the surface can be higher allowing it to carry a larger charge.

It now remains to show that a particle charged to the above calculated maximum can be driven against a gradient of 30,000 volts per cm. This can be done by making use of some data obtained by Deutsch[4] on the motion of charged particles in electric fields in connection with a study of the Cottrell process of precipitating dust from gases.

He found that particles of radius $r=10^{-4}$, after having picked up a charge of 376 electrons $=1.8\times10^{-7}$ e.s.u. in a corona discharge moved with a velocity of 0.56 cm/sec. in a field of 300 volts/cm. Particles of $r=10^{-5}$ cm picked up a charge of 5×10^{-9} and moved with a velocity of 0.42 cm/sec. in the same field. If we assume that the mobility varies linearly with the charge on a particle, we can use these results to determine the velocity of a particle of $r=10^{-4}$ cm and carrying the maximum charge (10^{-6} e.s.u.) in a field of 300 volts/cm. The velocity will be $v=(0.56\times10^{-6})/(1.8\times10^{-7})=3$ cm per sec. For the case of particles of radius 10^{-5} the velocity is less than 0.42 because the calculated maximum charge tuins out to be less than that observed by Deutsch. With the above velocity of 3 cm/sec. in a field of 300 volts/cm, a particle of $r=10^{-4}$, carrying a charge of 10^{-6} placed in a field of 30,000 volts/cm would move with a velocity of 300 cm/sec. Now a particle of this size can easily be blown with a velocity ten times as great. Apparently there is no difficulty to be expected in blowing the charged particles away from a highly charged conductor.

EXPERIMENTAL

The following experimental method was used to find the powder most suitable for the purpose in view. Fig. 2 shows the experimental arrangement.

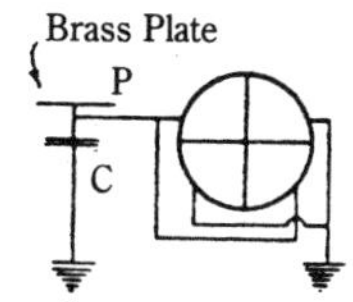

Fig. 2.—Experimental arrangement for investigating powers for charges.

An insulated brass plate P was connected to one pair of quadrants of a Dolezalek electrometer and to a condenser c, as shown. One milligram of the powdered material to be investigated was placed on the brass plate, which was earthed and insulated before blowing off the powder with a puff of air. The magnitude of the charge produced was determined from the deflection of the electrometer and the capacity of the system. The powders were prepared by grinding various solids and sifting them through a a 300-mesh sieve. This could not be done very well with metals which were used as obtained in the form of considerably coarser powders. The materials studied were mercuric

[4] Deutsch, Ann. d. Physik **4**, 824 (1930).

5-4 Continued

sulfide, mercuric iodide, sulfur, rosin, iron powder, antimony powder, clay, and diatomaceous earth, which is a form of silica occurring naturally in a very finely divided form.

The most promising materials were the metal powders and the diatomaceous earth. The metal powders could not be further investigated by the next method to be described because, owing to their great density, they could not readily be blown by the compressed air available. The diatomaceous earth turned out to be ideal for the purpose, not only because it gave such large charges, but also because it is very light and easily blown. It consists of particles 10^{-4} cm in diameter and smaller, and it can be obtained commercially at 50 dollars a ton.

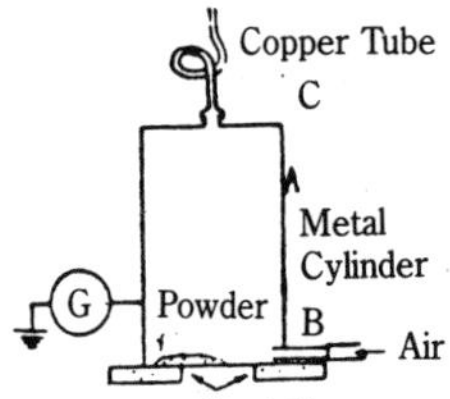

Fig. 3.—Experimental arrangement for measuring charge obtainable from diatomaceous earth.

The diatomaceous earth was used in larger quantities in such a manner as to permit the charges produced to be measured on a galvanometer. It was placed in a metal cylinder *A*, Fig. 3, 12 cm diameter and 30 cm high from which it was blown by means of compressed air introduced tangentially at *B*. The air laden with the powder passed through a piece of copper tubing *C* having an inside diameter of 0.5 cm and a length of 20 cm. The cylinder, insulated by standing on blocks of paraffin, was connected to ground through a calibrated galvanometer which indicated the current flowing from the cylinder as the powder was being blown out. The measurements were made by placing 5 grams of the powder in the cylinder and blowing it out with air flowing at the rate of one liter per second. The current is read on the galvanometer until all the powder is gone. The current is a maximum at the beginning of a run and decreases gradually as the amount of powder blown out per second decreases during the progress of the run. Since the current fluctuated somewhat, an average current was estimated during each half minute interval. These averages are listed in the second column of Table I which gives the result of a typical run. It should be noted that the average current recorded for the first interval is really too low because the initial swinging of the galvanometer prevents the current from being read at all during the first 15 seconds, during which time the current is considerably higher. A small current could still be read after the air had passed for 15 minutes. This is due to the fact that a small amount of the powder clung to the inner surface of the cylinder from which it was gradually dislodged and blown out by the air. The charge obtained per gram of powder is for these reasons somewhat higher than that given at the end of the table. The total charge in coulombs obtained from 5

TABLE I. *Charge obtained from 5 grams of diatomaceous earth blown by air flowing at rate of 1 liter/sec.*

Time in min.	Current in amp. $\times 10^8$	Charge in coulombs $\times 10^8$	Time in min.	Current in amp. $\times 10^8$	Charge in coulombs $\times 10^8$
0.5	137.5	4125	8.0	16.5	495
1.0	96.3	2889	8.5	13.8	414
1.5	68.8	2064	9.0	10.5	315
2.0	55.0	1650	9.5	8.3	249
2.5	41.3	1239	10.0	5.5	165
3.0	41.3	1239	10.5	4.7	141
3.5	41.3	1239	11.0	3.9	117
4.0	57.8	1734	11.5	3.0	90
4.5	82.5	2475	12.0	2.8	84
5.0	82.5	2475	12.5	2.2	66
5.5	68.8	2064	13.0	1.9	57
6.0	63.3	1899	13.5	1.9	57
6.5	46.8	1404	14.0	1.4	42
7.0	35.8	1074	14.5	0.8	24
7.5	27.5	825			

Total $30{,}696 \times 10^{-8}$ coulombs
or 6.1×10^{-5} coulombs/gram

grams of powder was found by adding together the product of current in amperes and time in seconds for all the half minute intervals. The charge on the powder is negative.

The copper tube *C* in Fig. 3 was at first straight, and it was found that the total charge obtained increased about 25 percent by bending it as shown. This is probably caused by an increased number of particles striking the wall of the tube due to centrifugal force on them. No further charge was obtained by either lengthening or shortening the tube.

The charge given by 5 grams of powder reaches the surprising value of 3.07×10^{-4} coulombs or 6.14×10^{-5} coulombs per gram. According to this value only 16 grams would have to be blown per second to get a current of 1 milliampere. Altogether 20 such runs were made, giving results which deviated at the most 11 percent from those given in Table I. None of the vagaries, such as reversal of sign, usually associated with frictional electricity were ever observed. A few runs made with lower air velocities gave much lower results, ranging from 3.02×10^{-5} coulombs per gram for the lowest air velocity capable of carrying the dust out of the cylinder and up. This is believed to be due to the cohering of the particles of the powder. The individual particles are approximately 10^{-4} cm in diameter and smaller, but they cling together forming larger aggregates which are blown apart by the air stream, the more completely the higher the velocity. It seems likely that larger charges per gram might be obtained by using higher air velocities, but this point could not be proved because the air pressure available was limited to two atmospheres.

The diatomaceous earth used to obtain the above results contained 12 percent of adsorbed water. No difference resulted by using the powder dried at 300°C for 1 hour.

5-4 Continued

A small scale model of a high voltage generator using diatomaceous earth blown by air was constructed as shown in Fig. 4. In this figure, *A* represents an insulated sphere of spun copper 20 cm in diameter, within which 8 copper tubes *B*, 0.5 cm inside diameter, were mounted. Compressed air introduced at *C* carried along with it diatomaceous earth introduced by a small screw conveyor *D* from a reservoir and blew it through the copper tubes. The charged powder left through a short length of glass tube and escaped into the air.

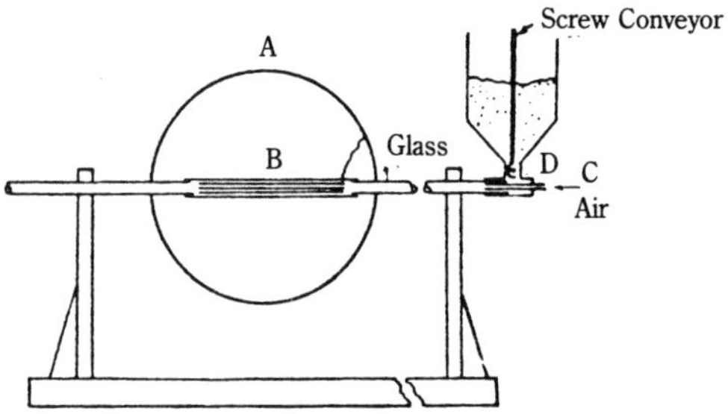

Fig. 4.—Experimental high voltage generator.

The potential of the sphere was estimated from the distance between it and a similar grounded sphere placed at such a distance from it that a thin spark jumped the air gap between them. The maximum potential reached seemed to be limited by a corona discharge from the sharp edges around the two openings in the sphere. The potential was estimated to be 260 kilovolts. A current of 8×10^{-5} amperes could be drawn from the sphere at this voltage by connecting it to ground through a glass tube filled with water and in series with a microammeter. The amount of powder introduced per second by the screw conveyor was 1.5 grams per sec.

In conclusion the writer wishes to thank Professor Millikan for very kindly placing the facilities of the California Institute of Technology at his disposal.

5-4 Continued

charging. Of course, in a double-disc generator, the distance between the two discs must also be kept small. These distances are difficult to obtain with thin plastic discs but somewhat easier to achieve with glass plates—which remain flat and have good temperature stability.

Precision construction with small clearances boosts efficiency in electrostatic generators just as it does in electromagnetic motors and generators. The magnetic flux circuit requires a small gap between the rotor and stator to maintain a strong magnetic field and to conserve the energy needed to produce that field.

Liquid, gas, and vacuum chambers

Another method for increasing the electric field strength across generator parts, extensively explored from the 1930s to the 1960s, is to completely enclose the generator in a chamber for experimenting with compressed gases, dielectric liquids, or vacuums. Even though the use of compressed gases in electrostatics was recommended as early as 1886 by the German scientist Hempel, his discovery was ignored until Robert Van de Graaff revived the idea in the early 1930s!

The use of dry compressed air, carbon dioxide, nitrogen, or hydrogen can greatly improve efficiency. An electrostatic generator in an environment of pure hydrogen at 15 to 20 atmospheres with an efficiency of 90% has been built. Even though this experiment occurred in the 1950s, there is still a widespread misconception that this type of generator is merely an inefficient novelty or toy.

Although this method of improving efficiency is quite exotic for those with limited shop facilities, one simple generator was insulated with compressed air. This invention received British patent number 22,731 in 1900 and the inventor was a Mr. Tudsbury. This generator was a small Wimshurst unit enclosed in an airtight metal case. The compression seal and packing nut provided for the hand crank extended through the case wall.

Tudsbury claimed that a generator with 8-inch-diameter plates produced sparks 2¼ inches long at normal air pressure, 6 inches long at 15 psi, and 9 inches long at 30 psi!

Again, the compressed air must be dried. It is also a good practice to include a drying agent, such as silica gel, within the chamber. Now that acrylic and *Lexan* plastics are on the market, the chamber could be made transparent for inspection purposes. If you are working along this line, you should design the unit for easy accessibility and generator adjustment.

You can see from this discussion, that many avenues are open for generator design. The results are limited only by your imagination. If even a small portion of the creative talent and money now invested in electromagnetic and semiconductor technology was applied to electrostatic generators, entirely new technologies could open up in the near future.

6

Theories of generator operation

IN THIS SECTION, TWO SAMPLES OF THE ORTHODOX, OR TEXTBOOK, explanations of how influence or induction generators work are included. It is fair to say that there are as many theories as there are inventors of original generator designs.

From the 1880s through the 1920s, heated debates appeared in the electricity journals concerning whose idea worked the best. The issue still has not been resolved; that is, no single theory explains all the properties that characterize these generators.

Because this is mainly a how-to book, I have included only the two most popular theories and limit the discussion to Wimshurst's disc-type generator (Figs. 6-1 and 6-2).

The first theory explains the mechanism as one of *influence*, or inductive action by one plate and its sectors on the opposite plate. The second theory avoids reliance on induction and is novel in its use of relative motion between charged bodies. There, the electric field is seen as stretching or relaxing, like a rubber band, between the plates and the neutralizing rods.

Presently, the influence machine is thought of as a continuously variable condenser; the discs form plates. In the relationship $V = Q/C$, where V is the voltage, Q is the charge in coulombs, and C is the capacitance in farads. If we can decrease the capacitance value when a fixed charge is given to a capacitor with movable plates, the voltage or potential difference across the plates will increase.

Machine with oppositely rotating Plates.—We now come to the type of machine which has been most recently introduced by Mr. Wimshurst, and which is more especially associated with his name.

It was first described in January 1883, and, as may be seen from the illustration (Fig. 62), differs essentially in construction from either of the machines which preceded it. It also differs greatly in its behaviour, for it is self-exciting, and will discharge its torrents of electricity under atmospheric conditions which are fatal to the working of the other forms of electrical machines. Moreover, the direction of the current will not change when the machine is at work; nor will the excitement die away when the terminals are opened beyond the limit of the sparking length.

Following the plan which we have already adopted, in dealing with other machines, we shall explain the action of the Wimshurst machine by means of a diagram (Fig. 61), before describing the different forms of it which have been constructed.

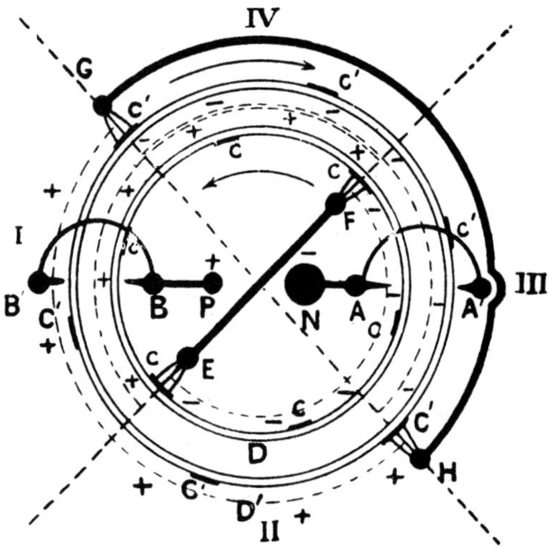

Fig. 61.—Diagram of Wimshurst Machine.

In the diagram, the oppositely rotating discs are represented by Bertin's method, as sections of cylinders D and D^1. The inner circle D represents the front disc, the outer circle D^1 the back disc, and each rotates in a

6-1 Conventional theory of Wimshurst machine. *Electrical Influence Machines*, John Gray, 1903

direction shown by the adjacent arrow. On each disc there is fixed a series of metallic carriers, denoted by the letters C and C^1. A neutralising rod E F is fixed on the disc D, so as to connect diametrically opposite carriers as they pass and touch the contact brushes fixed on its extreme ends; and a similar neutralising rod G H is mounted on the face of the disc D^1, with its contact brushes in a diameter at right angles to E F. The electrodes B P and A N have discharging knobs P and N at one end, and collecting combs B B^1 and A A^1 at the other. The collecting combs B B^1 stand facing each other, one in front of each disc, and are metallically connected by a bent rod passing round the edges of the discs; A and A^1 are fitted up and connected in a precisely similar manner.

This machine, like all those having metallic carriers, is self-exciting, that is to say, it requires no charge of electricity imparted to it from an external source to set it in action. The action of the machine does not depend on the presence of the collecting combs; for the charges on the surfaces of the plates are produced though the collecting combs are removed. In considering the development of electricity in the machine, the collecting combs and the electrode circuit may be left out of account. It is not yet certainly known how the initial charge is produced which leads to the starting of self-exciting machines, but probably it is due to the fact that there are no two places in the atmosphere at exactly the same potential at any given moment. As a consequence of this, we shall suppose that one half G B^1 H of the disc D^1 has a small positive charge, and the other half H A^1 G has a small negative charge. Under the influence of the positive charge on D^1, each carrier C on D, when in contact with the brush E, will receive a negative charge, which will be carried round by the motion of the disc till every carrier on the half E A F of the disc has a negative charge. In the same way, the influence of the negative charge on D^1 will impart a positive charge to each carrier C as it passes under the brush F, by which operation the half F B E of the disc D will be coated with positive electricity. The two halves into which the disc is divided by its neutralising rod have thus far acted

6-1 Continued

as field plates, to induce charges on the two halves into which the disc D is divided by its neutralising rod. The two halves of D now in turn act as field plates to induce charges on the carriers C^1 of the disc D^1, and a little consideration will show that, owing to the motion of this latter disc, the charges originally upon it will be increased by this action. The discs will continually react upon each other in the manner just described, and raise the potential of each other's charges according to the compound interest law, till the limit fixed by leakage is reached. The final distribution reached, is shown by the dotted lines, and + and − signs on the diagram. If we consider the discs divided into four quadrants (I, II, III, IV), by the two neutralising rods, it will be found that the charges on the two discs in quadrants II and IV are opposite, and, therefore, attract each other, but on the two quadrants (I and III) they are the same, and therefore repel each other. The collecting combs are placed opposite the middle of the quadrants (I and III) to draw off the self-repelling electricities. This, as shown in the diagram, imparts a positive charge to the discharging knob P, and a negative charge to the discharging knob N, and if these knobs are not too far apart, sparks will pass between them. As the action of the machine takes place without the collecting combs, it is not necessary that they should be in contact in order to start the machine. It is for the same reason impossible to reverse the action of this machine by separating the discharging knobs beyond the striking distance.

6-1 Continued

Collector combs and Leyden jars can be seen as accessories that are not needed for the generator to begin charging. Wimshurst did make a simple machine, with only the two plates and two pairs of neutralizer rods, which worked well. The metal sectors are not needed, and the generator is more efficient without them. The sectors mainly provide the convenience of self-starting.

Sectorless influence machines, such as I prefer to make, can in a sense be viewed as having two plates with an "infinite" number of tiny sectors. The surface particles, which compose the discs, become the charge carriers—the charge leakage is thereby reduced.

647. *New Theory of the Wimshurst Machine.* **F. V. Dwelshauvers-Dery.** (Deutsch. Phys. Gesell., Verh. 4. 13. pp. 276–277, Sept. 23, 1902. Paper read before the 74 Naturforscherversammlung at Carlsbad.)—At the request of de Heen the author contributed the following results: If an insulated body AB is moved rapidly towards a charged body C it becomes similarly charged (see Fig. 1). This phenomenon cannot be due to induction, since the charges at A and B are similar. If the motion is in the opposite direction so that the conductors are separated the sign of the charge in AB is reversed. These experimental results suggest the following explanation of the Wimshurst machine, in which the second plate is indicated by dotted lines and dashed letters. Suppose that there is a trace of positive charge in C (Fig. 3). The elements of the sector bOC are approaching C; hence they become positively charged, and the charge in C is increased. In the sector

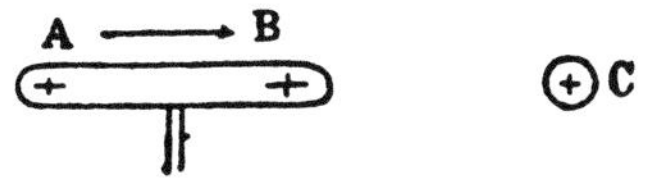

Fig. 1.

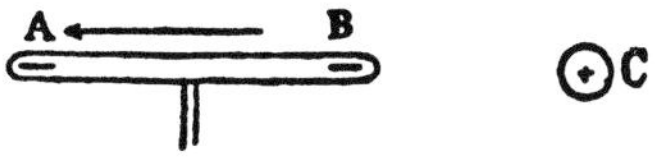

Fig. 2.

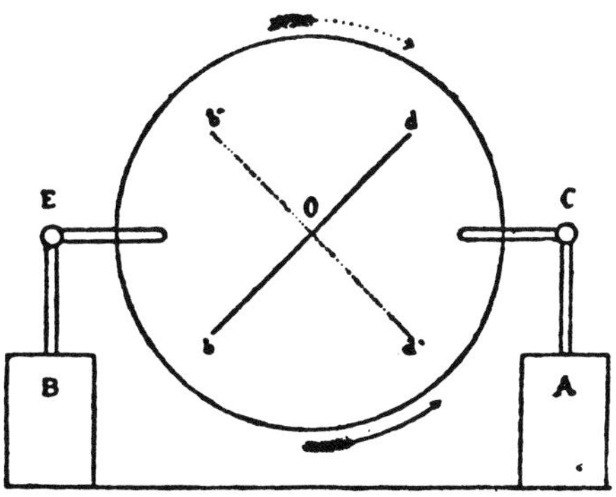

Fig. 3.

COd the reverse takes place, though the change is numerically less since the surface is less. The elements of the sector dOE become negatively charged since it is receding from C, hence E becomes negatively charged. A trace of electricity in either C or E thus determines the poles of the machine. In the above only one plate has been dealt with. The machine does give charges when only one plate is used, but they are so small that an electroscope is required for their detection. When both plates are used the second acts like the first, the positively charged sector b′OC increases the charge in C, and the negatively charged sector increases the charge in E. The most important action, however, is that of the one plate on the other. In the angle dOd′, for instance, the positively charged elements are approaching one another, and hence become more and more charged. This action is greatest when the distance between the plates is as small as possible. J. E.-M.

6-2 An alternative theory. *Science Abstracts*, A, 1903

Generally speaking, there are a number of generator properties, which an adequate theory would need, to explain the following characteristics:

- The source of the initial charges in self-starting generators with sectors. Both contact potential and cosmic rays have been invoked.
- The buildup of potential and what determines the upper *voltage* limit, other than corona leakage.
- Sudden polarity reversal, which was mostly a problem with the Holtz and Varley generators. Wimshurst machines generally have a stable voltage output.
- The importance of the disc surface—its microstructure, semiconducting or insulating properties, and surface charge distribution.
- The role of the surrounding medium, whether it is dry air, compressed gas, dielectric liquid, or vacuum. Why is pure compressed hydrogen gas very effective in spite of its low dielectric strength?
- The generator power sometimes spontaneously dies out when the discharge terminals are separated beyond the maximum striking distance.
- The sensitiveness of generator performance to neutralizer system design and its circuit resistivity.
- The disc material decomposes with time. In the case of acrylic plastic, the plates become coated with fine white dust after extended periods of operation. The conventional explanation is that generator discs are decomposed by ozone gas.

These are the major characteristics that have appeared after many hours of experimentation. Of fundamental importance is the very nature of *electrification*, whether by friction, impact, contact, or induction. I will discuss the subject of electrification later in chapter 9.

2
PART
Accessory instruments, experiments, and applications

In the second section of this book, I will describe some of the basic electrical devices that can be used for extensive exploration into the nature of electrification. Although the cost of the sectorless influence machine that I describe might range from $100 to $200, depending on the tools available in your shop, the electroscope, Leyden jars, and the electrophorus are much less expensive. These latter projects require only simple and inexpensive tools for their construction. However, do not assume, as most textbooks on electricity would lead you to believe that these instruments are well understood in operational theory.

Following the electrophorus discussion, I will finish with some of the many unusual experiments in high-voltage electricity that are rarely mentioned in classroom physics books. You will see that there are a number of unexplored avenues awaiting the pioneer in electrical science.

7
The electroscope

THE *ELECTROSCOPE* IS A DEVICE THAT DETERMINES THE ELECTRICAL condition of its surrounding atmosphere; specifically, it determines the presence of charges on nearby bodies or the polarity (+ or −) of each charge. The basic electroscopic instrument does not display an absolute numerical measure; instead a simple scale is provided, graduated in degrees. As in Fig. 7-1, a typical design consists of a metal rod, on top of which is mounted a ball or disc, and carrying at its lower either a single strip of gold leaf and a metal plate or two strips of gold leaf. The rod and leaf are electrically well insulated from ground, and the delicate leaves are protected from air currents. To establish that a body is electrified, bring it close to the top terminal. If the leaves diverge, electricity is present. Normally, a glass or plastic rod rubbed with a dry cloth is used to charge the scope.

Figure 7-2 shows some of the possible electrified states that the electroscope can exhibit. The description of electroscopic charging in Fig. 7-2, written in 1875, is the same explanation that would be given today in physics textbooks.

Later in the text are experiment results that suggest that the classical explanations are superficial and based on too few experimental tests. First, however, I will describe how you can build your own scope using common materials. A source is provided for the sulfur insulation for the leaf (see Appendix A).

The sensitive gold leaf is sold through picture-frame or craft stores in book form with about 25 leaves. Each leaf is about 5 inches square and separated by tissue paper. Each gold leaf is only about 0.1 micron thick (about four-millionths of an inch)!

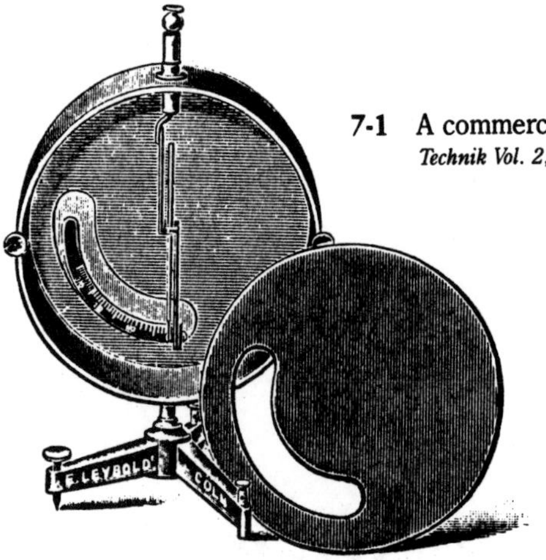

7-1 A commercial electroscope. ***Physikalische Technik** Vol. 2*, J. Frick, 1907

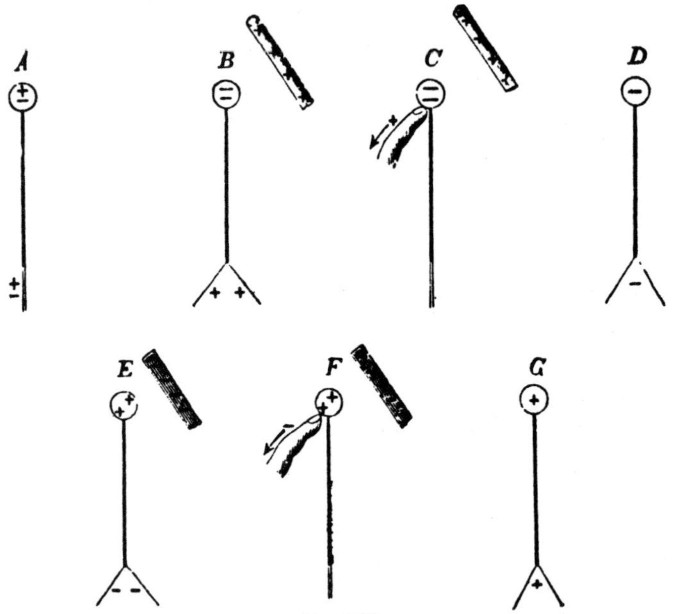

FIG. 308.

In the ordinary state the electroscope contains equal quantities of positive and negative electricity, as represented at *A* in fig. 308. When a body charged, say, with positive electricity, is brought near it, electrical separation is effected, the negative electricity being

7-2 Various charged states of electroscopes. ***Introduction to Experimental Physics***, **Adolf Weinhold**, 1875

attracted to the knob, the positive electricity being repelled to the further extremity, that is, to the gold leaves. These are now electric, and being both charged with the same electricity, they repel each other, that is, being flexible, they *diverge*, as at *B*. When the inducing body is removed, the leaves drop again because the two separated electricities recombine, and the electroscope is again in the neutral state *A*. If the metal rod is touched with the finger, while the inducing body is still near and the two electricities therefore still separated, then, as shown at *C*, the free positive electricity of the gold leaves escapes through the finger and the leaves drop, while the negative electricity of the knob remains bound as long as the inducing body is near the instrument. Now, while the inducing body is still near, let the finger be removed first, and next the inducing body; then the negative electricity becomes free, and as it cannot now escape, it diffuses itself over the metal portion of the instrument, and the gold leaves diverge again, as in *D*.

The electroscope is now charged by induction with the opposite electricity to that of the inducing body, that is, negatively, if the inducing body was, for example, a positively charged glass rod, and positively if the inducing body had been a negatively charged stick of sealing-wax. The latter case is represented in figures *E*, *F*, and *G*, which correspond to figures *B*, *C*, and *D*, respectively.

An electroscope, *charged* with either electricity, may be used not only for deciding whether a body is in the electric state or not, but also with what kind of electricity the body is charged. When a neutral body is brought into the neighbourhood of a charged electroscope as shown in *A* and *B* of fig. 309, no appreciable change takes place in the divergence of the leaves. In reality the divergence does diminish in that case slightly, because the electroscope itself acts now like an inducing body, separating the electricities of the neutral body brought near it, and in consequence of mutual inductive action, a quantity of electricity of the opposite kind to that with which the instrument is charged, is repelled to the leaves, neutralising a portion of the original charge.

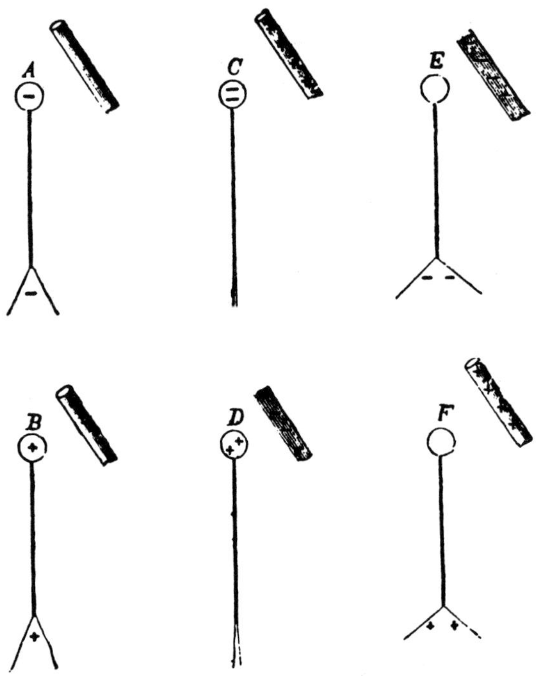

Fig. 309.

This action is, however, so slight, compared with the effects of a charged body, that no mistake can be made in the conclusions. If a body charged with opposite electricity is brought near to the instrument, as in *C* and *D*, the leaves drop because their charge is attracted to the knob. If a body charged with the same electricity is brought near, the charge in the knob is repelled to the leaves, and their divergence increases as in *E* and *F*.

The real gold leaf books are expensive, but if you are in a high school science class, you might share the cost of one book with several students. Otherwise, try composition gold leaf or aluminum leaf, which are quite inexpensive. As a last resort, use the thinnest aluminum foil from the grocery store. Gold leaf is much more sensitive because it is so thin and has a very low mass for its size.

Building a cookie-tin electroscope

Once you have purchased your leaf book, get a cookie container such as is often used for packing fancy imported cookies. These metal containers measure about $3^1/_2$ inches in depth and $7^1/_2$ inches in diameter.

Older physics books usually picture electroscopes housed in glass bottles or flasks. These housings are not acceptable because the nonmetallic surfaces near the leaf might become charged and, in some cases, attract and tear it.

To reduce the influence of extraneous electric fields, it is best to surround the gold leaf with a metallic surface provided with small windows for visibility. The metallic enclosure acts as a barrier to electric fields so that even if a high charge exists on the shield's outer surface, the inner surface remains neutrally charged. Such an enclosure is called a *Faraday cylinder*. The cylinder should also be grounded for best results.

The completely assembled cookie-tin electroscope is shown in Fig. 7-3. This homemade inexpensive leaf electroscope with graduated scale can be used to measure high resistances, atmospheric electrical tensions, and radioactive samples. Figure 7-4 shows the interior of the electroscope.

Directions for building the cookie-tin electroscope

Clean the cookie tin with soap and water, then dry.

In the front and back, cut out a $2^1/_2$-inch-diameter opening and remove the burrs for a smooth edge.

Into the top of the cylinder, cut out a 1-inch-diameter hole for the cork and smooth the edges. If you have a thin rubber grommet available, you can use it to line the hole; it looks more attractive.

Saw the wood base to measure $^3/_4 \times 2 \times 8$ inches and cut a saddle shape in the top surface for mounting the cylinder. Use a faucet bevel washer, flat-head machine screw, and nut to secure the cylinder to the base. Cut 3-inch-diameter windows from sheet mica or use glass lens covers (for flashlights) and glue them inside the $2^1/_2$-inch-diameter holes, front and back. Put a thin film of glycerine on the outer surfaces of these two windows to make them conductive.

7-3 The cookie tin electroscope.

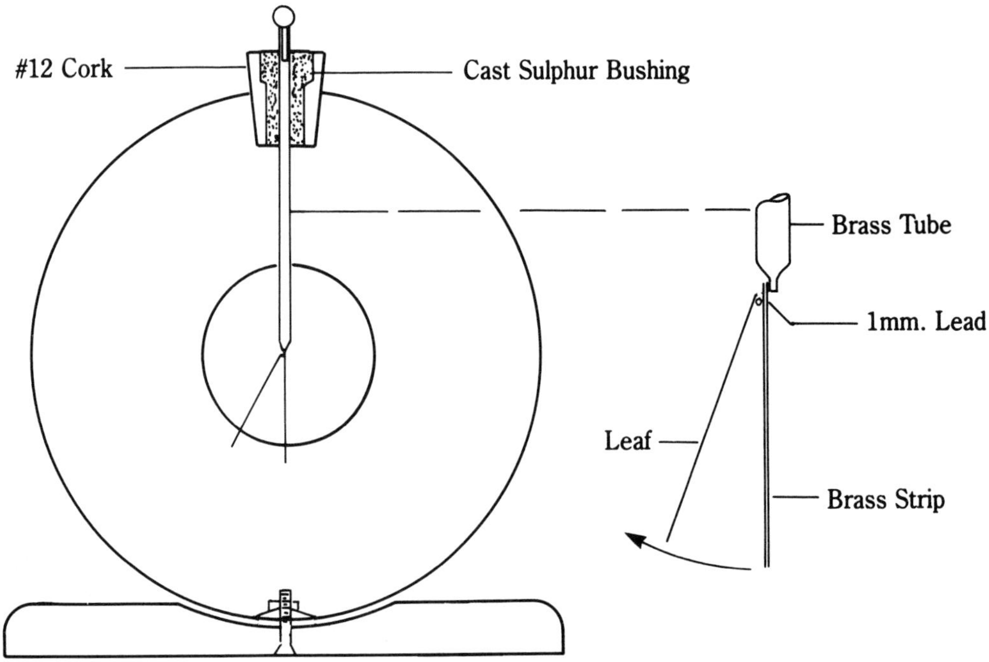

7-4 A construction sketch for the electroscope.

The leaf support is a thin wall brass tube 1/8 inch i.d. and 4 1/2 inches long. Cut a strip of brass 1/32 × 1/4 × 1 5/8 inches, round its ends, and polish well to remove all sharp edges. Solder this strip on the flattened bottom of the tube, as shown. Glue a 1/4-inch length of 1-mm-diameter pencil lead horizontally across the back of the brass strip to provide the point of support for the gold leaf.

Next, cut a #12 cork with a stepped through-hole, one size of which is 3/4 inch in diameter and the other, 5/8 inch in diameter, provide good clearance from the brass tube in the center.

Cover the bottom of the cork with adhesive aluminum foil, making sure it is well sealed around the bottom and side of the cork. Using a pointed tapered rod, pierce a 1/8-inch hole in the center of the tape and carefully enlarge it so that the brass stem (1/8-inch i.d.) just slides through. Find the desired position between the tube and the top of the cork by inserting the cork in the 1-inch hole of the cookie tin. The 1-mm lead (see Fig. 7-4) must be centered in the window, since it is the pivotal point for the gold leaf. When you find the best position, mark it.

Now, lightly hold the brass tube vertically in the vise and rest the finished cork on top of the vise jaws in the desired position. Note that the hole in the aluminum tape fits tightly around the brass stem and that the stem is concentric with the hole in the cork.

Place chunks of lump sulfur (brimstone) in a small porcelain crucible and cover with the lid. *Caution*: sulfur should not be exposed to fire, since it burns readily. Heat the crucible very gently over a low flame—overheating destroys the insulating property of sulfur. The molten sulfur will be orange-yellow.

Remove the flame and pour the liquid sulfur slowly into your #12 cork. Fill until the sulfur is level with the cork's top. Allow the bushing to cool slowly and avoid causing vibrations while the casting sets. After about 5 minutes, the sulfur will be crystallized, and this medium yellow in color. When it is hard, but still warm, peel off the foil seal around the cork's base. You now have a custom-cast bushing made of sulfur—one of the best insulators! After one day, the sulfur will assume a lemon-yellow color if it has not been overheated.

Sulfur holds charges much better than Teflon™ and styrene plastics—two otherwise good insulators. For your own test purposes, when the leaf is charged to a deflection of 80 degrees (almost horizontal), it requires two hours to drop 15 degrees, which is the 65 degree position (temperature 64°F, relative humidity 45 percent).

When assembling the gold leaf, work in a draft-free room and cover your mouth and nose to avoid blowing on the fragile leaf. Using a new single-edge razor blade, remove one sheet of gold leaf and its two protective tissue papers, and place this sandwich on flat cardboard. With a sharp pencil, sketch out a section $1/4 \times 1\,1/2$ inch on the top paper tissue. Draw the blade over this section gently cutting through the back sheet. Using two clean needles, remove the gold leaf and lay it on the cardboard. Now lay a strip of tissue paper across the top of the pencil-lead support. Using the needles, work the gold leaf onto the brass strip in the desired position, with the leaf's top edge resting on the tissue and the support lead underneath.

Lay a second tissue strip over the leaf body for protection and place a small weight on it to prevent movement. Lift the first tissue off the lead support, moisten the lead with a trace of saliva, remove the same tissue, and press the upper edge of your leaf onto its support. It will dry in a minute. Next, remove the weight and carefully remove the last cover tissue. Slowly raise the leaf support system and cork, and position them in the 1-inch hole at top. Solder the two removable terminals—a ball and a 1-inch-diameter disc—to 1/8-inch-diameter brass rod; these terminals slide into the top of the brass tube.

Be careful not to overcharge the leaf and tear it. Its maximum deflection should be about 90 degrees (the horizontal position).

To protect the insulating value of your electroscope, cover the cork and terminal with a metal cap. This measure prevents dust from settling on the sulfur bushing.

The electroscope was especially useful at the turn of the century. Then, it was used to study radioactivity and the natural ionization of the air. Both Ernest Rutherford and C.T.R. Wilson studied and designed very sensitive gold-leaf electroscopes. Rutherford's book *Radio-activity* describes a modified electroscope (see Fig. 7-5).

Rutherford calculated the current sensitivity of an electroscope with a 1-liter volume and a 4-cm-long gold leaf and rod support system. The electric capacity C is usually 1 electrostatic unit. If V is the decrease of potential of the leaf system in t seconds, the current i through the 1-liter volume of air is:

$$i = \frac{CV}{t}$$

Since the fall of potential in air averages 6 volts per hour, then:

$$i = \frac{1 \times 6}{3600 \times 300} = 5.6 \times 10^{-6} \textit{ electrostatic units}$$

Therefore, $i = 1.9 \times 10^{-15}$ amperes.

56. A modified form of the gold-leaf electroscope can be used to determine extraordinarily minute currents with accuracy, and can be employed in cases where a sensitive electrometer is unable to detect the current. A special type of electroscope has been used by Elster and Geitel, in their experiments on the natural ionization of the atmosphere. A very convenient type of electroscope to measure the current due to minute ionization of the gas is shown in Fig. 12.

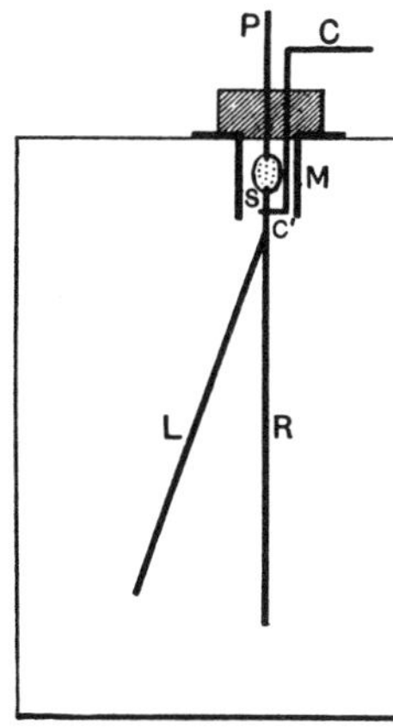

Fig. 12.

This type of instrument was first used by C. T. R. Wilson* in his experiments of the natural ionization of air in closed vessels. A brass cylindrical vessel is taken of about 1 litre capacity. The gold-leaf system, consisting of a narrow strip of gold-leaf *L* attached to a flat rod *R*, is insulated inside the vessel by the small sulphur bead or piece of amber *S*, supported from the rod *P*. In a dry atmosphere a clean sulphur bead or piece of amber is almost a perfect insulator. The system is charged by a light bent rod *CC'* passing through an ebonite cork. The rod *C* is connected to one terminal of a battery of small accumulators of 200 to 300 volts. If these are absent, the system can be charged by means of a rod of sealing-wax. The charging rod *CC'* is then removed from contact with the gold-leaf system. The rods *P* and *C* and the cylinder are then connected with earth.

The rate of movement of the gold-leaf is observed by a reading microscope through two holes in the cylinder, covered with thin mica. In cases where the natural ionization due to the enclosed air in the cylinder is to be measured accurately, it is advisable to enclose the supporting and charging rod and sulphur bead inside a small metal cylinder *M* connected to earth, so that only the charged gold-leaf system is exposed in the main volume of the air.

In an apparatus of this kind the small leakage over the sulphur bead can be eliminated almost completely by keeping the rod *P* charged to the average potential of the gold-leaf system during the observation. This method has been used with great success by C. T. R. Wilson (*loc. cit.*). Such refinements, however, are generally unnecessary, except in investigations of the natural ionization of gases at low pressures, when the conduction leak over the sulphur bead is comparable with the discharge due to the ionized gas.

* Wilson, *Proc. Roy. Soc.* Vol. 68, p. 152, 1901.

7-5 Rutherford's sensitive electroscope. *Radio-activity,* E. Rutherford, 1905

The amazing current sensitivity of such a simple scientific instrument clearly illustrates that you don't always need exotic and expensive equipment to do careful research. But, much practice and manipulative skill is needed to produce these delicate devices. C.T.R. Wilson developed the skill of cutting gold leaf to such a degree that he could produce leaf strips 1/10 mm in width! You must try cutting gold leaf to appreciate this feat.

A scale for registering leaf deflection is made with tracing paper, on which increments from 0 (vertical) to 90 degrees (horizontal) are marked. The paper scale is positioned behind the outside back window; a light shining behind the scale illuminates it for easy reading.

It is especially interesting to lecture to large classes with the electroscope placed between the condenser lens and compound objective lens of a projector or "magic lantern." The light source will project a shadow of the leaves onto a large screen marked with a scale. This way, the movement can be shown to many students at once (see Fig. 7-6). Typically, the condenser lens has a focal length of 10 cm, and the objective lens (for focusing the image) has a focal length of 5 cm.

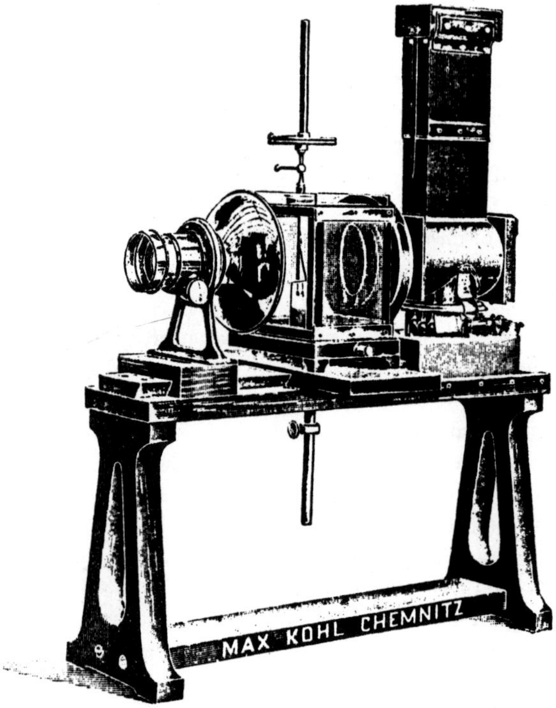

7-6 Condensing electroscope with lantern projector. *Physikalische Technik Vol. 2*, J. Frick, 1907

Electroscope anomalies

Most physics texts describe the electroscope as basically an "ionization chamber." This description is given because the charge leakage from the system is said to be the result of ions in the air within the electroscopic chamber. In 1901, C.T.R. Wilson claimed that the leaf discharge rate is the same during day and night; therefore, leakage is not caused by light. Sir William Crookes found that when electroscopes are evacuated, leakage is greatly reduced, and in general, leakage is proportional to atmospheric pressure.

Unusual fluctuations in leaves, immediately after charging, are said to be caused by unequal heating of the air in the chamber from incident light beams. To determine if the orthodox explanation is justified, you can repeat Wilson's and Crookes' experiments.

One experiment that must be performed in clear, sunny weather, preferably outdoors, with a relative humidity of less than 20%, involves charging the scope to a deflection of about 50 degrees. After charging to this degree of deflection, remove the charging rod. Properly done, the leaf will increase its deflection spontaneously over the next few minutes by up to 20 degrees! It would appear that the energy is flowing uphill, from low to high, as though the charged scope is seeking a new level of energy equilibrium with its environment.

Now charge the electroscope by induction. The leaf then has a charge opposite to that of the charging rod. Slowly bring the rod from a distance toward the scope terminal. The leaf deflection decreases as you bring the rod closer. But as you draw the rod even closer, the leaf which is now at about zero deflection, begins to rise again. Finally, the leaves reach a high deflection when the rod is very close!

Is there an energy "node" around the instrument, with different properties on either side of the node? Can you think of a setup with more than one node about the electroscope?

Other anomalous electroscopic experiments vary the geometry of the materials making up the device and the types of materials themselves.

The quintessential, but unfortunately uncelebrated, electroscopic experiments were performed by the amateur experimenter, Dr. Gustave Le Bon of Belgium. His two books on physics, *The Evolution of Forces* (1908) and *The Evolution of Matter* (1910), caused an uproar in scientific circles. I assume the uproar was because the works were so original and thought-provoking.

Figure 7-7 shows the electroscopic ball surrounded by three concentric metallic screens of varying thicknesses resting on an insulated plate. Under the influence of the charged horizontal rod above, the leaves diverge. Conventional theory says that electric forces should not penetrate the screens, since the screens act as Faraday cages.

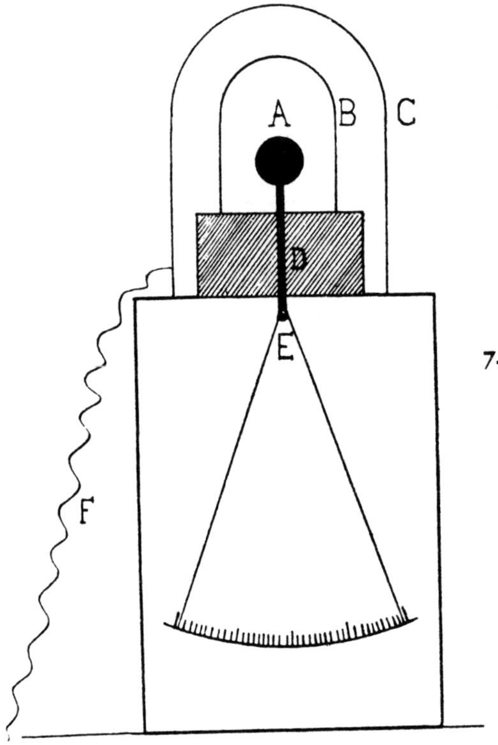

7-7 A screened electroscope. *The Evolution of Matter*, Gustave Le Bon, 1907.

In Fig. 7-8 are three electroscopes, each with a different terminal shape over which is supported a negatively charged rod. Especially unusual is that the charged rod is kept positioned for some time, up to 12 minutes, in order for the leaves to be affected. When the rods are finally removed from the electroscopes, each will have a different electrified state—neutral, positive, or negative. Why is time an important factor?

In another experiment, a radioactive substance in a metal capsule is placed on the plate of the electroscope, and a metallic plate, such as aluminum, is positioned above on a stand. You might not expect the metal blade to influence the discharge rate, if the increased leakage is caused by radioactive ionization. The type of metal plate and its condition are also important. The same peculiar result was independently verified by physicists William Ramsay and W.S. Lazarus-Barlow about 1905.

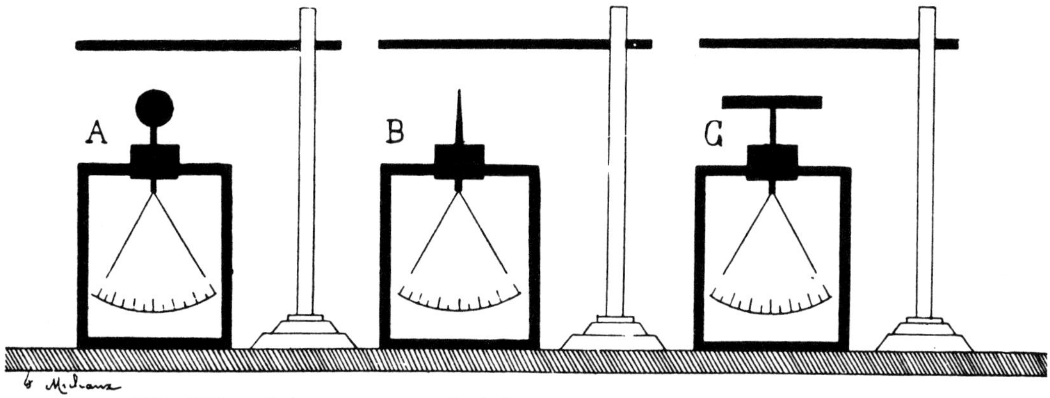

7-8 Effect of electroscope terminal shapes. *The Evolution of Forces*, Gustave Le Bon, 1908

The last of Le Bon's ingenious experiments I mention concerns electroscopic tests comparing the dissociation of metals by sunlight and radioactive substances (see Fig. 7-9). Because Le Bon's experimental conclusions were so disturbing and seemed so unorthodox to the scientists of his day and because he did not happen to possess "credentials" as a physicist, his contributions were not acknowledged in most journals. He was probably the first scientist to recognize and announce the universal and general dissociation of all matter. Many other physicists went on to build careers and receive awards for work that had beginnings in the research of this great scientific experimenter. Such is the true history of innovation!

Research avenues

In addition to using electroscopes to study the nature of electrification and the resistivity of insulators and semiconductors, you will soon realize that the instrument is very sensitive to weather changes.

Maximum leaf deflection and slow leakage occur when the sky is bright and clear, and the humidity is low. Minimum deflection and fast decay occur during humid, overcast days, in heavy fogs, and after heavy rains.

Several questions might occur at this point: How does the scope behave before and after thunderstorms? How is leaf deflection related to relative humidity? How is leaf deflection related to barometric pressure and temperature?

By placing a well-insulated bare wire horizontal to the ground and joining it to your scope, you can detect a buildup of charge, especially in clear sunny weather. How does the charge vary with the height above the ground? Does the type of metal make a difference? Is the sun radiating charged particles that electrify the wire?

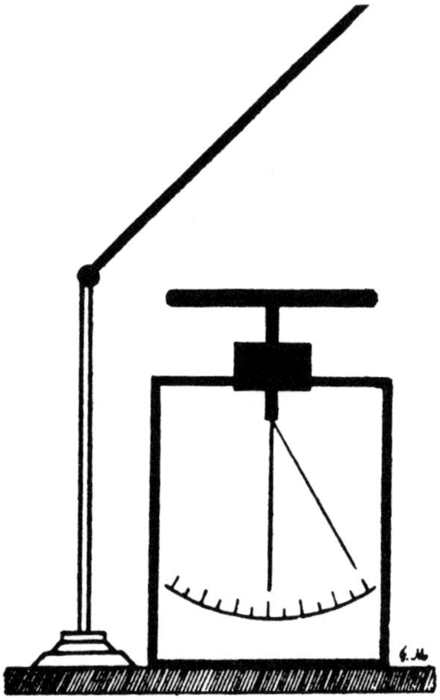

7-9 Influence of sunlight on metals.
Evolution of Matter

Fig. 9.—*Comparison of the dissociation of spontaneously radio-active bodies and of metals under the influence of light.*—A tin mirror prepared as described in the text and a screen of the same size coated with oxide of thorium or of uranium are used alternately. The dissociation of the atoms of the tin under the influence of light is forty times more rapid than that of the radio-active bodies just mentioned.

Do electroscopes charge spontaneously during earthquakes? They have been shown to do so. An excellent book, which devotes a chapter to this subject, is titled *When the Snakes Awake—Animals and Earthquake Prediction* (1982) by Helmut Tributsch.

The connection between earthquakes and electrometers was first mentioned in 1799 in Venezuela. The world traveler, Alexander Von Humboldt, wrote that his electrometer oscillated oddly just before two earth tremors. Later, in 1808 the Italian scientist, Vassalli-Eandi, found the atmospheric electricity too large to measure during a tremor.

The weather during a quake is often described as being very still and oppressively heavy. Perhaps the piezoelectric or triboelectric effects cause an increase in charge.

How does the electroscope behave before the onset of a tornado? The electrical properties of tornadoes have been carefully documented in *Lightning, Auroras and Nocturnal Lights* (1982) by William Corliss. I will discuss homemade electric tornadoes in a later chapter.

I have devoted considerable space to the electroscope in this book because its importance to basic research is seldom mentioned in physics texts. Its simple appearance is quite deceptive. Although electronic circuits are now available for measuring charges, they cannot deliver the useful information given from study of leaf motions and the rates of leaf fall.

When all the anomalous experiments—usually omitted from classroom discussions—are taken into account, they point toward a need for a simpler, less-contrived explanation. The *mechanistic theory,* which relies on static forces, seems incomplete; what is needed is a *dynamic theory.*

Remember, when designing electroscopes, to surround the leaf system with a grounded metallic shield and to keep the dimensions of the leaf, leaf-support, and terminal small for maximum sensitivity to a given charge.

8
The Leyden jar condenser

THE LEYDEN JAR IS ESSENTIALLY A FLAT PLATE CAPACITOR: ITS METALLIC coatings and dielectric are rolled up in cylindrical form. Figure 8-1 shows one of the earlier forms of the jar in use during the late 1700s.

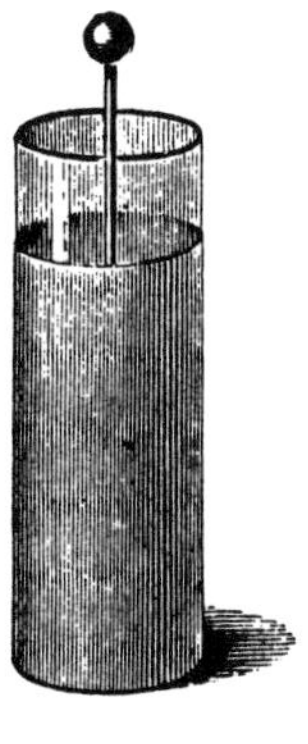

8-1 Lidless Leyden jar condenser.
Physikalische Technik Vol. 2, Frick, 1907

The Leyden jar is recorded as being first developed by Ewald von Kleist in 1745. In 1746, Pieter van Musschenbroeck of Leyden, Holland, further experimented with the invention. The jar's original form consisted of a nail immersed in a bottle partly filled with water. Instead of using a bottle with an outside coating, Musschenbroeck merely wrapped his hand around the bottle. The jar was more or less grounded through the operator's feet. Good grounding of the outside coating to earth was found to be essential when charging the Leyden jar.

A very good discussion of the history of the Leyden jar's development, including the early theories about its action, can be found in *Electricity in the 17th and 18th Centuries* (1979) by J. L. Heilbron.

The most interesting facet of this experiment worth emphasizing to the electrical hobbyist is that the jar was not discovered by following the accepted rules of electricity of the day, but rather by being unaware of them. This fact of history provides good ''food for thought,'' even today. We generally think that worldwide communication assists scientific innovation, yet I believe that incomplete communication can, and often has, allowed an innovator to think and try an experiment that those steeped in orthodoxy refuse to consider. Complete communication does foster a bland uniformity, and it cannot always give impetus to creative developments.

Dielectric constants

The amount of energy, which is stored in a condenser, is expressed in the formula:

$$E = 1/2\, CV^2$$

C, the electrical capacitance, increases with the area of the jar coatings, with the kind of insulator, or *dielectric*, and also with the thickness of the insulation separating the coatings.

V is the *applied voltage*, or *potential difference*, that is produced between the two coatings.

Each dielectric is assigned a number that varies with the internal structure of the material. That number is termed the *dielectric constant*. Some typical dielectric constants are given in Table 8-1. The higher the number, the greater the capacitance, and therefore, the larger the stored energy.

Table 8-1.
Dielectric constants for insulators.

Insulator	Dielectric Constant
Plate Glass	6.8 - 8.4
Pyrex Glass	4.1 - 6.1
Hard Rubber	3.0
Polystyrene	2.5
Shellac Film	4.0
Spar Varnish	4.8 - 5.5
Bees wax (purified)	2.9
Paraffin wax	2.5
Wood (dry)	2.0 - 5.2
Air	1.0

So, a Leyden jar made from a glass bottle will produce a hotter spark than one using a plastic bottle of the same dimensions. Since many types of glass attract moisture at room temperature, you should first test its ability to hold a charge before actually building a condenser.

When choosing jars, flint, crown, and Pyrex glass are recommended. The jars should have openings large enough to admit your hand (for applying the inner coating), and the glass should be clear and bubble-free. To reduce electrical stresses, the jar should be cylindrical and have rounded edges at the bottom.

Once you select the jar, clean and dry it, finish with a lint-free cloth dipped in alcohol to remove grease.

Warm the jar in an oven to about 100°F, remove, and rub the side briskly with a silk or satin cloth. Bring this rubbed area of the glass near the pole of your electroscope to test for the presence of a charge. If the charge is favorable, you should coat the jars, while warm, with varnish inside and out; shellac varnish, polyurethane, or copal varnish is recommended. These finishes prevent leakage from moisture. Only Pyrex glass jars need no coating of varnish because they are not hygroscopic. Cure the jar in a dry, dust-free warm room.

In the discussion of my generator, I gave details for building Leyden jars with metal powders and filings as the conductive coating. These powders and filings scintillate when charging and do leak slowly over time. Actually, this leakage is a safety factor because it reduces the shock hazard.

A more leak-resistant condenser might employ brass or steel shim stock, 0.002 to 0.004 inch thick, or as a last resort, heavy-duty aluminum foil. Varnish thinly and evenly applied to the inside of the jar will secure foil coating, but epoxy glue is better for the stiffer shim stock. Always place factory-cut edges at the top side of the jar to reduce leakage. Keep the shim stock in place with wood splints or rubber bands to prevent separation. The charge leakage can be reduced further by applying melted paraffin to the top edges of both coatings.

Lids, which support the inner metal pole of the jar, were originally made from cork or baked mahogany dipped in a paraffin bath. Finally, the bottoms of the jars are either left uncoated or sprinkled with metal filings inside and covered with foil on the outside. Metal filings stick best when you first coat the glass with Elmer's wood glue. In Fig. 8-2 is a general design for a "dry" condenser, which uses metal coatings.

Many jars can be joined together in parallel to form a Leyden "battery," but time is saved by resorting to the "wet" condenser which uses conducting liquids in place of metal coatings. The jars should, of course, be varnished as before. It is important to keep the liquid levels at the

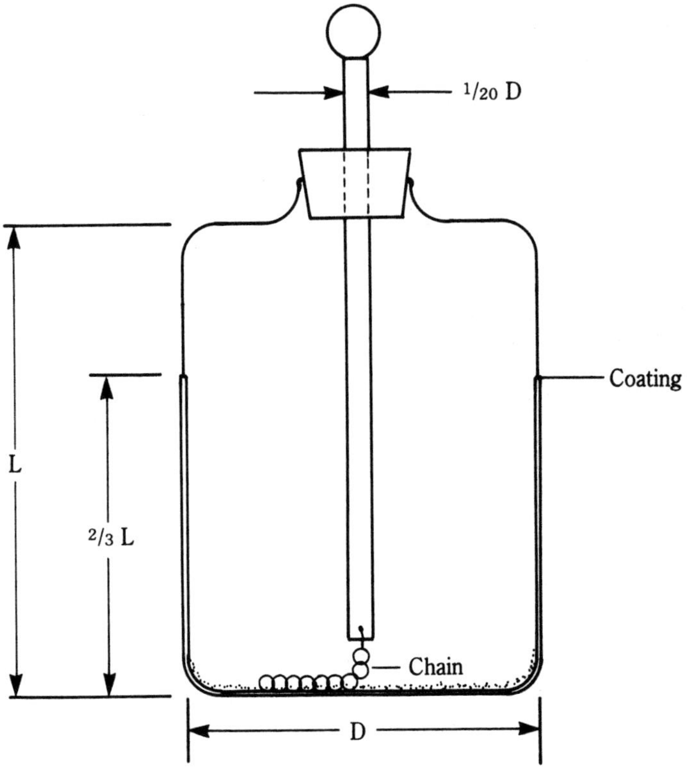

8-2 A well-proportioned Leyden jar.

same height, both inside and outside, to avoid electrical stress. A description is provided in Fig. 8-3.

Although a layer of linseed oil or castor oil can be used for insulating liquid condensers, I recommend paraffin oil, which has better electrical properties.

Like the electroscope, the condenser is a very important tool for research. Even though it appears to be simple and straightforward in design, a careful investigation of it can provide a wealth of knowledge about stored energy and the nature of electrification in dielectrics.

Design modifications of Leyden jars

One of the least-explored areas of the Leyden jar involves making subtle changes in its materials and geometry. I have already indicated experiments with the electroscope that involve jar design changes. It follows that it is possible to vary the quality of the electricity and the behavior of the jar.

A Liquid Condenser

In the condenser described below, instead of using tinfoil for coating the dielectric, a conducting liquid covered with ½ inch of boiled linseed oil is employed, a practice which obviates all brush discharge. It also has the additional advantage of insuring perfect contact between dielectric and coating, thus obviating all uneven distribution of strain and decreasing the liability of a breakdown.

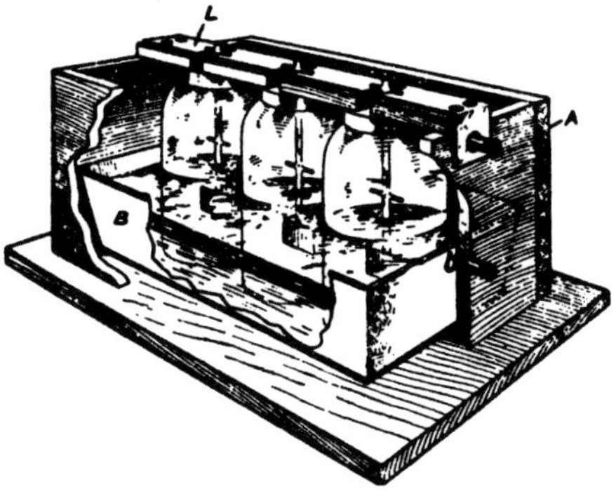

LIQUID CONDENSER

A box (A) of proper dimensions is constructed of paraffined oak. The strip top (L) should have deep notches sawed in it to correspond with the mouths of the jars. A bread pan (B), long enough to hold three jars side by side, is placed in the box and in it three Mason pint jars.

The conducting liquid is then poured to within ¾ inch of the upper edge of bread pan and over it is placed ½ inch of boiled linseed oil. The jars are filled in the same manner. The conducting liquid may be either dilute sulphuric acid (fifteen parts water and one part acid), a strong solution of sal ammoniac, or a salt water solution.

Three connectors are constructed by soldering a short length of No. 8 copper wire to a copper plate. Two supports of No. 12 copper wire are soldered at right angles to the large wire and are so spaced that when the upper one rests on the connecting strip the plate is immersed in the conducting liquid, and when the lower one rests on the connecting strip the plate clears the liquid by ½ inch or more.

The top is then put in place and over it the connecting strip. The three connectors are then slipped in place and the top and connecting strip are screwed to the box. A binding post is mounted on the bent end of the connecting strip. Another binding post is mounted on the same end of the box and fastened to it is a No. 12 wire which makes contact with the bread pan.

The capacity of one of these units of three jars is .00234 microfarads when all the jars are connected, .00156 when two are connected, and .00078 microfarads when one jar is connected, for it is a well known principle that when a number of condensers are connected in parallel the total capacity is equal to the sum of the individual capacities of the several condensers.

8-3 Liquid condensers in parallel. *Popular Electricity*, 1912

One of the recent tendencies of scientific research is what might be considered a drift toward increasing sterility and uniformity. We picture the scientists in clean white suits working in a windowless laboratory, well isolated and insulated from the natural environment. This tendency appears to be our collective philosophy—almost an unconcious attitude of how we view science as a mental discipline.

In the eighteenth and nineteenth centuries, a different wind was blowing in science. There were no specialists in the front lines of research, but rather amateur investigators, broadly educated and highly skilled, whose main desire was to satisfy their intellectual curiosity. They were not called ***physicists***, but rather ***natural philosophers***.

We would do well to revive the spirit of inquiry from those earlier days. Do not be afraid to cross over into disciplines outside of or seemingly separate from, physics.

For example, when exploring creative design concepts involving the Leyden jar condenser, you might break from the idea that the coatings must be perfect metallic conductors in the form of uniformly thin sheets. Consider not only coatings made with metal powders, turnings, or shot, but also intermetallic compounds and semiconducting stone powders. In like manner, consider materials other than ideal insulators, like glass or plastic. For example, try semiconductors that become insulators at elevated temperatures. This area was explored from 1759 to 1762 by Edward Delaval of Cambridge, England. He found that stone, clay, and charred wood, when heated, lose their moisture and become good insulators.

When doing creative work you should get out and explore in the natural setting for inspiring ideas. You are working with natural energies in this branch of electrical science, and these same energies shape and form the environment. It is obvious that natural conditions are neither sterile nor uniform. Note the shapes of sea shells, insect bodies, seeds, etc.

The demands of contemporary physics show that entropy can only increase—meaning that the universe is running downhill. When we consider the energetics of biological and weather systems, we see abundant evidence of a run-up potential leading to increasing order. The new science of chaos is a study of the increasing order in natural systems. Both increasing entropy and decreasing entropy are needed to give a full picture of the equilibrium-seeking natural energies in the environment.

Finally, I give a word of warning. A one-pint Leyden jar, fully charged, packs a wallop sufficient to knock you off your feet. Show a charged large jar (one to five gallons) the same respect you would give to any powerful explosive. Use the special discharging tongs, for safety's sake. Never use wires or screwdrivers (see Figs. 8-4 and 8-5). Finally, store the jars with jumper wires joining the inside and outside coatings. This method ensures that residual charges can't build up, as often happens when using glass as a dielectric.

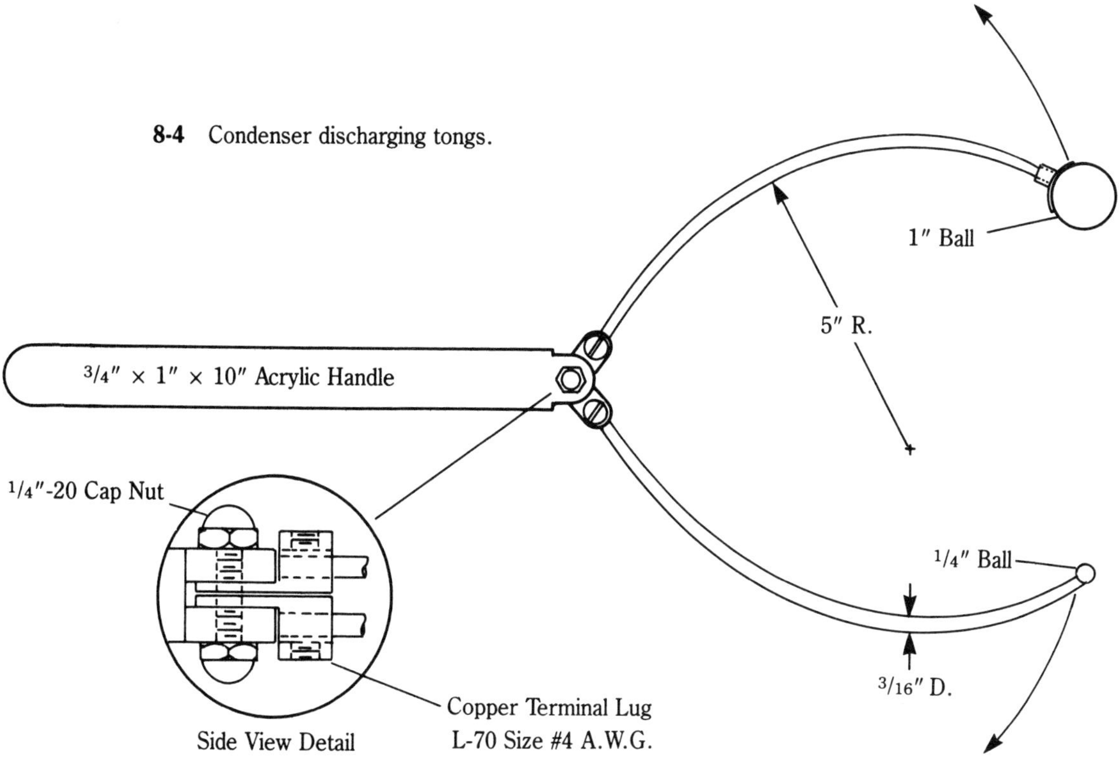

8-4 Condenser discharging tongs.

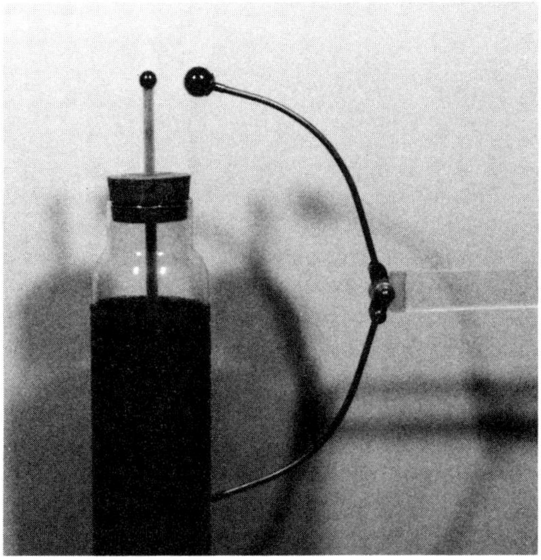

8-5 Proper way to discharge Leyden jars.

9
The electrophorus

THE LAST BASIC INSTRUMENT DISCUSSED IN THIS BOOK IS ESSENTIALLY A modified condenser with dissectable parts. When the coatings and dielectric are manipulated in a special way, the condenser becomes a charge-dispensing device, called an *electrophorus* (see Fig. 9-1).

Johannes Wilcke, who worked with dissectable condensers, developed this device in 1762. The device was later called the *perpetual electrophorus* because once the dielectric was charged, a seemingly endless quantity of electric charge was produced by special manipulation. In 1775, Alessandro Volta named the device and perfected its manipulation. His theory of its action was not as clear as Wilcke's explanation, however (see Fig. 9-2).

In Fig. 9-2:

A. The dielectric cake is charged by rubbing it briskly with a cloth. The cake is resting on a metal base that remains grounded.

B. The metallic cover is placed on the cake.

C. The like-induced charge is conducted to the ground. Then the connection is broken.

D. The cover with the opposite charge is lifted from the cake.

Since the cake is an insulator, it can't give all its charge to the cover or to the base by conduction, so its electrification, once produced, can persist for a long time. To operate the electrophorus, discharge the cover, replace it on the cake, temporarily ground it, and then remove it.

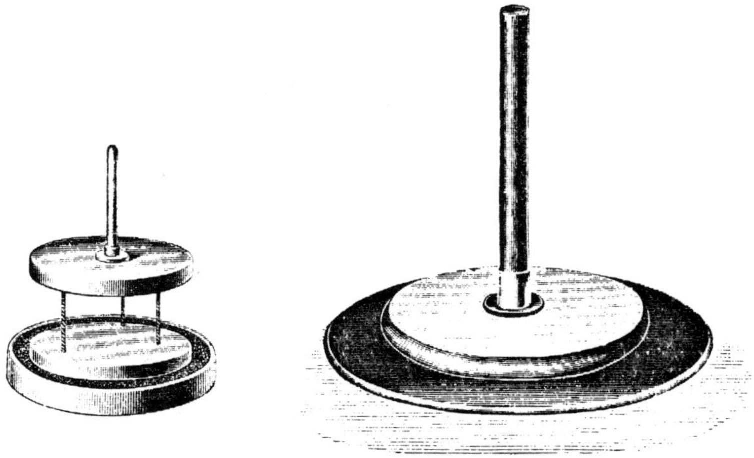

9-1 The basic electrophorus. *Phsikalische Technik* V.2, Frick, 1907

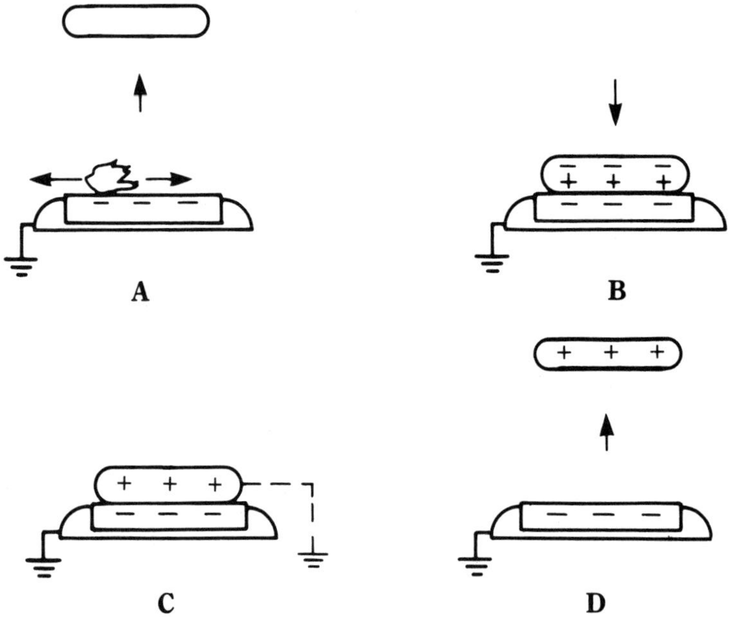

9-2 Steps in charging the electrophorus.

Operated properly, a continual charge supply is available, hence the name *perpetual electrophorus*.

Presently, the action is said to be caused by electric induction across surfaces in close proximity. Extra mechanical work is required to remove the cover because of its electrical adherence to the cake. Not well understood is just how the resin cake, an insulator, can continue to supply charges over a long period of time.

Note that the electrophorus is the electrical analog to a permanent magnet. The "keeper," placed on the magnet, preserves its power just as the cover plate preserves the cake's charge. Cakes have been known to preserve an initial charge for up to a year!

The grounded electrophorus

The first simple modification to the electrophorus was mentioned by J. Phillips in the *Philosophical Magazine* in 1833. Until that time, the experimenter's finger was used to temporarily ground the cover (see Fig. 9-2, STEP C). Phillips had the bright idea of pasting a thin, narrow strip of tin foil across the entire face of the cake, each end in contact with the metal sole plate. The foil automatically took the place of finger grounding, but the funny thing is that the foil short-circuited the plates of his dissectable condenser! Even so, the resin cake did not discharge.

Another method Phillips mentioned was to bore a small hole through the cake's center and place a metal plug in the hole, contacting the sole and flush with cake's top surface. Either method must allow intimate contact between the cover and the cake.

Electrical shock from a sheet of paper

One modified form of the electrophorus uses thin film semiconducting surfaces in place of thick resin cakes.

Get a metal japanned tea tray with rounded edges and a flat bottom. Then support it on two cakes of paraffin or dry glass tumblers. Cut a piece of heavy brown wrapping paper so that it covers the tray's bottom. Near the fireplace, hold the paper until it is quite warm, but not scorched. Hold one end down on a dry wood table, while giving it a dozen strokes with a latex rubber glove or neoprene balloon skin. Now, bring the sheet by two corners over the tray. It will fall like a stone. If you touch the tray, a noticeable shock will be felt. Note that this experiment must be done in a warm, dry room for best results.

What is interesting here is that by heating paper above room temperature, you can measurably increase its resistance by driving moisture from its pores.

Experiments with the thin film electrophorus were extensively covered by Professor Joseph Weber of Ingolstadt, Germany. His main work on this subject *Abhandlung von dem Luftelektrophor* (1779), contains a wealth of information. Unfortunately, it is written in the old German script, which makes translation difficult. His simple and inexpensive designs employed fabrics, such as linen or silk, tightly stretched on wooden frames. These frames were placed in the vertical plane near the fireplace. Once warm and dry, considerable amounts of electricity could be drawn off after the membranes were briskly rubbed. It is not clear if the stretched fabrics were first varnished after the silk was oiled.

Making the traditional cake electrophorus

The original electrophorus designs used dry glass handles for holding the cover plates, which must have well-rounded edges. To size the cover plate, make its diameter about 8/10 diameter of the resin cake.

When the electrician could not afford the expense of having a metal cover made, a fine-grained hardwood, such as mahogany, was shaped and planed smooth, and metal foil was pasted on it. The resin cake (dielectric) must be dense, smooth, and hard, and never cast less than 1/2 inch thick. Usually the thickness is 5 percent of the diameter of the cake.

Both the cover and the cake's mating surfaces must be perfectly flat—when pressed together, a slight vacuum forms.

As far as electrical output is concerned, a resin cake 8 inches in diameter can deliver about 1 microcoulomb of charge to its cover at each contact. The spark length is normally one-tenth the diameter of the cake. For example, an electrophorus cake 24 inches in diameter will produce a 2 1/2-inch spark. Here again, the device should be used in a warm, dry room for best results.

Baking a cake

Although brimstone (as molten sulfur) makes the best cake, it often cracks from temperature changes. Repeated liquefaction before casting and slow annealing after casting offset the cake's brittleness to some extent.

An electrician, William Snow Harris, provided several formulas for resin cakes, one of which follows:

Refined shellac flakes	1 part (by weight)
Venice turpentine	1 part
Resin	1 part

Melt shellac and resin slowly in a covered iron crock placed on a sand heat. When melting commences, add the turpentine and continue stirring with a glass rod. Once the solution is perfectly fluid, add a little coloring, such as ivory black, to produce a nice appearance. Retain the melt in this state until the air bubbles are expelled.

Caution: Never melt this combination over an open flame!

The casting surface was often polished marble or plate glass. You can try an acrylic plate as well. I recommend covering the casting surface with a very smooth sheet of aluminum foil to permit easy separation from the cake when cold.

Place a ring of sheet metal or wood of the desired thickness on the foil and place lead weights on the ring's edge to prevent the mixture from seeping out. Pour the melt into the mold, until the ring is filled and level. Allow this mixture to remain undisturbed until the cake is quite cold—overnight is best. If the wood ring must be removed from the cake, use a split ring, bound with a string and lined with foil, inside.

I prefer wooden embroidery hoops for the casting form; wooden basket hoops are also serviceable. Harris did not specify the type of resin used, but either ester gum or rosin would be a good choice. The melt temperature is normally 250°F, and this fluid condition should be maintained for at least 30 to 60 minutes.

When the cake is cold and the foil face is removed, the surface next to the foil, which is perfectly flat, becomes the top surface next to the cover plate. The cake can be placed in a metal cake pan, which serves as the bottom or sole plate. The pan should be electrically well grounded.

Furs (and wool) were originally used to excite the cake surface; but a sheet of silk, satin, latex rubber, or flannel will work. All materials must be warm and dry. Using oblique glancing blows sometimes works better than simply rubbing the surface. Now, holding the cover's handle, place it down on the cake, ground it temporarily, as described before, and lift the cover.

Bring your finger near the cover and you will promptly receive a sharp, snappy jolt—a miniature lightning bolt! Use Phillip's method for automatic grounding and you can draw a spark each time you lift the cover (without charging the plate). You can continue this process until you are tired. Always store your electrophorus with its cover on, to protect it from dust.

The smallest cakes have been only 1 inch in diameter (for charging electroscopes). One of the largest electrophorus cakes ever recorded was made in 1777 by Professor Lichtenberg. This monster was 6 Paris feet in diameter, and the cover had to be applied with a rope and pulley hung overhead. The cake consisted of resin, turpentine, and Burgundy pitch. The spark was said to be 15 inches in length—thick and hot.

Today, the study of the electrification of dielectric bodies centers around *electrets*. The main difference between electrets and resin cakes is that electrets are charged by applying a voltage while they are hardening. No mechanical rubbing of the surface is then required. A typical formula for an electret consists of 45% carnauba wax, 45% rosin, and 10% beeswax. These substances are melted together, as before mentioned.

The author's electrophorus

My generator is based on the electrician Tiberius Cavallo's design. Cavallo made a small 6-inch-diameter electrophorus using simply a thick glass plate with wax thinly melted on top. Sealing wax of "second quality" was recommended; this wax contains a small amount of magnesium carbonate chalk as a filler material.

One wax formula calls for (by weight):

Shellac flakes	4 parts
Resin	1 part
Venice turpentine	1 part
Powdered magnesium carbonate	1 part

Melt shellac and resin together in a copper pan on a hot plate, then add the turpentine and chalk slowly, stirring well. Once cooled, reduce this compound to powder by placing small chunks of it in a *pulverizing bowl*. This container is a large wood or metal bowl with steep sides, supported from overhead by three ropes. Place a 20-pound steel or lead ball in the bowl. Move the bowl in a circular motion to reduce the wax to powder. This setup is today called a *ball mill*.

Place the glass plate, well cleaned and warm, on wood blocks and uniformly sprinkle the top of the plate with the powdered sealing wax to a depth of about 1/16 inch. Place the plate and block support in an oven heated to just melt the wax, about 250°F. When the wax is fused and covers the glass plate, remove it and cool gradually. Trim wax from the edges with a warm knife. For best results as an electrophorus plate, the film of wax must be thin and flat. Adding turpentine helps to produce a thin flowing wax. This description is the traditional Cavallo method for producing waxed glass plates.

I prepared the sealing wax from this recipe:

Carnauba wax	8 oz.
Ester gum	2 oz.
Turpentine	4 oz.
Magnesium carbonate (powder)	1/2 oz.

Caution: This mixture is flammable! I recommend working outdoors with a fire extinguisher that is good for oil fires, nearby.

Heat the wax and estergum in a heavy aluminum bowl or skillet. Sift in the magnesium carbonate, stirring to eliminate all lumps. Use an old egg beater to mix the wax uniformly. Remove the container from the heat, pour in turpentine, and stir well. Pour the wax batter on aluminum foil to make a thin sheet. When the mixture is hard, place the wax and foil in a freezer for one hour. Then remove the foil and break the very brittle sheet into small chunks for the ball mill. I used a large aluminum bowl and a 10-pound lead ball, cast from two half-spheres. Refreeze the wax chunks, as needed, to keep the wax brittle for easier pulverizing.

Now apply the wax powder to a clean, warm, window-glass plate, 1/4 inch thick and 14 inches in diameter. Place the glass plate in a large metal pan, such as a pizza pan, and situate it over a hot plate. Carefully level the glass plate with a bubble level and gently heat the plate to about 150°F to remove moisture. Sift wax powder over the glass to a uniform depth of 1/16 inch. Gradually raise the temperature to 250°F, until the melted wax uniformly covers the glass. The thickness of the film should average 1/100 inch. Slowly lower the temperature over a one-hour interval to avoid thermal stresses. Use a steel rule to test for the flatness of the wax film.

Shown in Fig. 9-3 is the finished Cavallo electrophorus with the heat source below. Figure 9-4 shows the mounting of the heat source and Figs. 9-5 and 9-6 give construction details for the wood base of the electrophorus.

Put a sheet of aluminum foil on the bottom surface of the wax-covered glass plate, called the *dielectric plate*. Recess the whole glass plate into a wooden frame and secure it with wood screws and finish washers around the edge. Use a 40-watt tubular appliance bulb to provide some heat and remove moisture from the glass plate. A temperature of about 110°F is ideal for electrification.

9-3 Homemade electrophorus with integral heat source.

Use a 14-inch steel ''deep'' pizza pan (see Fig. 9-7) from a restaurant equipment supplier (aluminum pans do not have flat bottoms). Always test these pans by laying a 12-inch steel rule across the bottom. The top edges of the pan must have a large rolled bead to prevent charge leakage. The brass nut is called a ***ballcock coupling nut*** in the plumber's trade. This nut is used for toilet hookups; use only brass metal so that it can be soldered onto the steel pan. Clean the inside center of the pan and the flange of the brass nut with fine sandpaper, then apply acid-core solder to the flange of the nut only. Apply flux to the center of the pan and place the nut in position. Now place the pizza pan on top of the vise jaws. On the flange nut, place a small steel plate to weigh the nut down and to transfer heat downward.

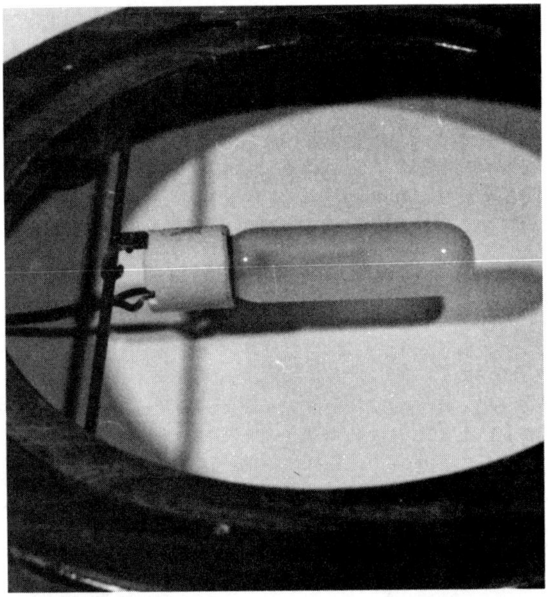

9-4 Heat source detail.

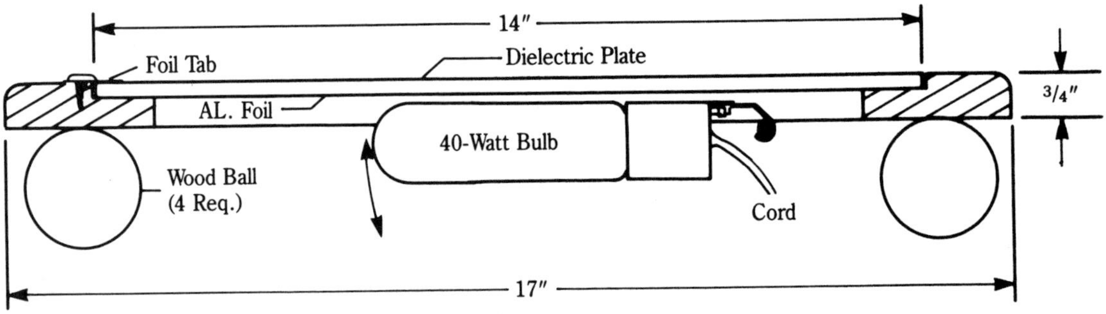

9-5 Side view.

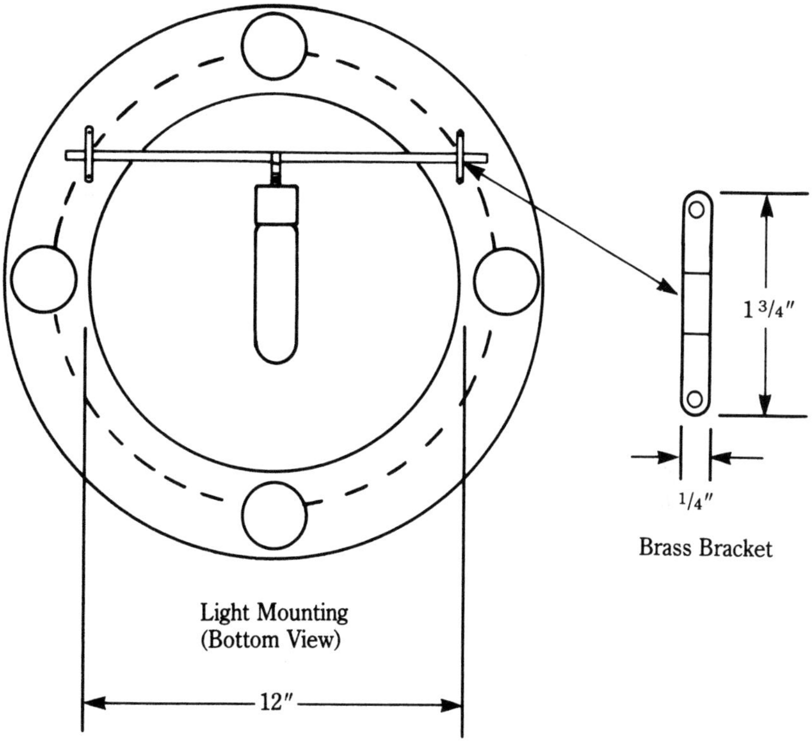

9-6 Bottom view.

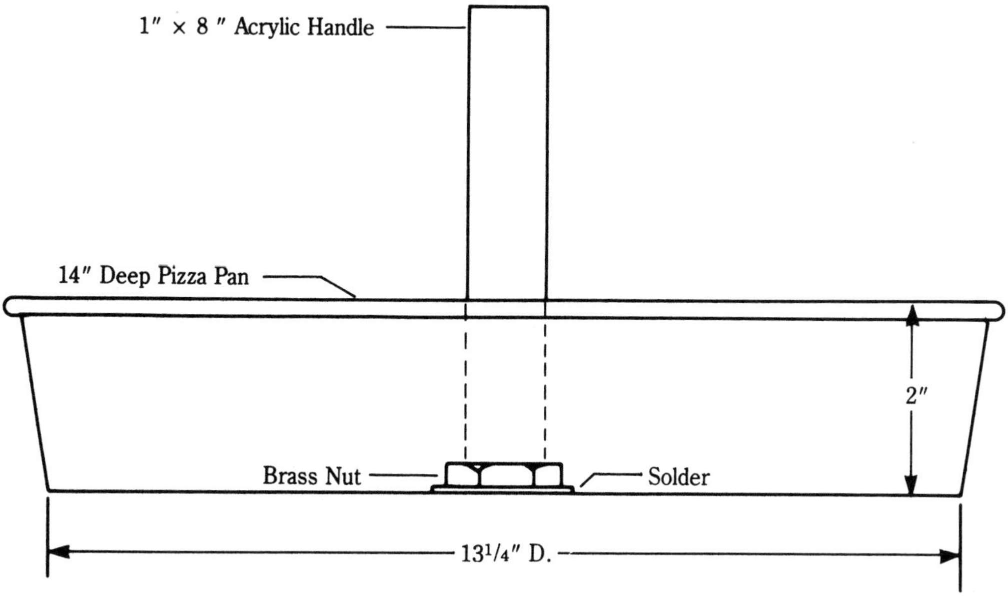

9-7 Metal cover and insulating handle.

Using a propane torch, direct the flame horizontally at this plate only. The heat will be transferred to the nut until the solder melts and settles onto the pan. *Caution*: Never heat the pizza pan directly. The heat will warp the bottom and make it useless as a cover for the electrophorus.

For the handle, use an acrylic rod or thick-walled tube 1 inch o.d. by 8 inches long. Thread this tube with a die, or in a lathe to about 15/16 inch in diameter by 14 threads per inch for a length of 1/2 inch on the bottom end. Then size to fit the brass nut snugly. Now, the electrophorus is completed.

Use the foil face below the dielectric plate (Fig. 9-5) to ground through the wood body of the device. You can run a jumper wire from the 1/4-inch brass rod passing through the two brackets (Fig. 9-6) to ground.

Briskly rub a large rubber balloon skin, surgical sheeting (known as *dental dam*), or surgical gloves over the warm dielectric to excite the waxed plate. Place the cover on top, ground it for a second, then lift it by its handle. If properly done, the plate will be charged and will be able to produce sparks of about 1 1/4 inches long. If you see leakage from the cover in the dark, look carefully for burrs around the upper bead of the pan and remove them with a fine file.

If you do not want to experiment with waxes or varnishes on glass plates, you can simply substitute a 1/4-inch acrylic disc with foil underneath and get fair results. I did try a wax electret formula on glass and found that it was too thick. This type of wax also develops stress cracks because turpentine is not used to aid in fusing the wax compound.

A simple charge indicator placed on the cover shows no charge when the cover is down and a high charge when the cover is raised. Figures 9-8 and 9-9 show "Judy," first at ground potential, then at 50,000 volts.

From the toy store, Judy came with blond plastic hair that I replaced with 1/8-inch strips of facial tissue paper. A thin strip of adhesive copper foil runs from her head to her feet to transfer the charge. The strands of paper are attached with wood glue. After her feet are glued to a metal base, Judy is transformed into a very pretty electroscope!

This experiment vividly illustrates the relationships among voltage, charge, and capacitance. Remember that the electrophorus is basically a dissectable flat plate capacitor with a metal cover, an insulator plate, and a grounded metal sole plate.

$$\text{Since, Voltage } (V) = \frac{\text{charge } (Q)}{\text{capacitance } (C)}$$

9-8 Folks, meet Judy!

9-9 Judy shows her potential.

For a given charge on the plates of the capacitor, as the cover plate is raised, the capacitance must decrease. This decrease in capacitance appears as an increase in voltage or potential difference between the top and bottom plate. The charge on the cover plate becomes unbounded, or *free*, and changes into a spark.

It might seem to those well grounded in basic physics that the electrophorus and its principle of operation are clearly understood; therefore, little is left for the electrical experimenter to improve upon. As I have mentioned before, there is a classical explanation based on the use of bound and unbound "static" charges that act on each other by induction (see Fig. 9-2).

Semiconductive stones

There is, however, a tiny "fly in the ointment," which appeared in a press release in *Popular Mechanics* in 1935. What goes on here? After I studied the article and its accompanying photo for some time, I guessed that a peculiar variation on the electrophorus was invented back then. The photograph showed the inventor with a wood and rubber "mushroom" that was claimed to produce an electrical charge. According to the article, a spiral of metal was in or on the electrically charged circular glass plate. The spiral's outer end could be grounded through the handle (see Fig. 9-10). A 12-inch-diameter wood "mushroom" device is covered over with a sheet of rubber. This convex shape and the membrane would make a very good means for exciting the plate of an electrophorus. The thick white block, which appears to be marble, looks like a sandwich of three pieces. Could there be a metal spiral inside it also? The article stated that when the block is "rubbed with the mushroom for five minutes, sufficient electricity is generated to keep a neon tube glowing for eight to twelve hours."

The article also stated that when a neon gas tube was moved back and forth in front of the plate, the tube would glow brightly.

A really strange question is this: If a neon light can be energized for several hours without rubbing the plate, of course, should we reconsider the theory of electrification, which relies on static charges? Does the electrophorus have a persistent dynamic energy flow that requires an initial rubbing action to get it started?

The electrical properties of semiconducting stones remain largely unexplored. However, accurate studies of these stones might be necessary in order to explain the operational theory of the electrophorus.

Alessandro Volta wrote a report in the *Philosophical Transactions*

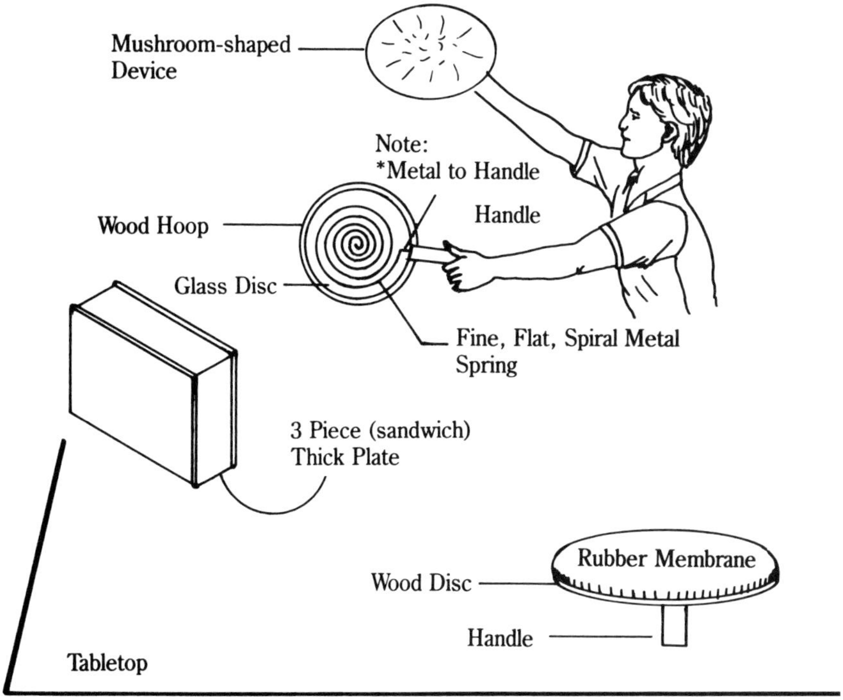

9-10 A charge-radiating electrophorus. Drawing based on *Popular Mechanics*, 1935

(1782) on his work with semiconducting materials. One discovery was that a conducting plate retains a charge better when it is in contact with a poorly insulated body (semiconductor) than when resting on a resin cake (which is well insulated)! This discovery became known as *Volta's Paradox*. One of the semiconducting materials best suited for this experiment was a polished plate of white Carrara marble, warmed up to drive off moisture.

Related to Volta's Paradox is the *Johnsen-Rahbek effect*, described in 1923. This discovery concerns the tendency for certain kinds of stone to adhere strongly to metal plates when a high-voltage direct current is applied. The adhesive force, from electrostatic attraction, was found to increase with the fifth power of the applied voltage. Several patents, such as 1,533,757, April 1925, were issued to Johnsen and Rahbek. They preferred stones that were finely porous and hygroscopic—slate, marble, steatite, agate, limestone, flint, and jasper.

The physics of electrical adhesion is not well understood. As far as the mushroom generator from Berlin is concerned, no U.S. patent was found, but there is a possibility that a German patent was issued. If so, it probably would have appeared between 1930 and 1939.

Some avenues for electrical pioneers

Although it is one of the simplest electrical generators, the electrophorus should be considered as important as the electroscope and the condenser to basic research on the nature of electrification. The long-persistent charged condition of the dielectric is still not well understood. Because of this general lack of understanding, I have gone to considerable lengths to explain the traditional methods of making this generator.

Several questions present themselves at this point:

- How are the plate's conditions (smoothness and thinness) important to electrification and its persistence?
- In semiconducting plates, how do surface resistance, bulk resistance, porosity, moisture content, and temperature affect electrification?
- What are the best materials and what is the best shape for the rubbing device used for charging the electrophorus's plate.
- Should the sole or bottom plate of the device be a solid sheet of metal or should it spiral from the center to the circumference? If it is to spiral, would an arithmetic or a logarithmic shape be best?
- How does the length of time spent rubbing the electrophorus affect the persistence of electrification?

When returning to nature for examples, note how often multiple layers are used in constructions. The layers in onions, tree trunks, rocks, minerals, and soils suggest experiments with layered electrophorus plates using alternating films of dielectric and semiconducting materials, for instance glass and wax. Experiments along these lines can help us to visualize what takes place during the condition we call *electrification*. Until this is visualized, we cannot really understand, and therefore cannot design to enhance this process.

A more complicated generator, such as the Wimshurst machine, will use the same basic principles found in the electrophorus. However, the simple must be understood before the complex can be explained.

Summary

The electrophorus, when it first appeared, caused almost as much consternation as its predecessor, the Leyden jar. As a result, some major theories were made or dashed.

In visualizing electrification, we might, for example, picture the plate of an electrophorus as covered with depressions. Each depression is occupied by a freely suspended marble. A glancing impact to the surface will set several of the marbles spinning like gyroscopes. This persistence of motion might serve as a dynamic model of what happens during electrification at the molecular level.

Although the costs of making an electroscope, Leyden jar, and electrophorus are quite reasonable, you will need to master some new skills in this work. I advise you to keep your notes in a journal, especially recording the daily weather conditions. The amount of knowledge gained with these instruments is limited only by your ability to create original experiments. As Ernest Rutherford said years ago, "An ounce of thought is worth a ton of equipment."

10
Electrostatic motors

MANY SUBJECTS IN HIGH-VOLTAGE WORK ARE GREAT SCIENCE FAIR AND college graduate research projects. Because of the subtle variations possible in the design and materials used in these experiments, a single avenue of information can provide a wealth of knowledge for the researcher over many years.

An excellent book that discusses the history, types, and principles of operation is *Electrostatic Motors* by Oleg Jefimenko (1973). This electrostatic motor book traces electrostatic motors from their beginnings in the eighteenth century up to about 1970. An excellent bibliography is included.

Conventional electric motors use electromagnetic force to convert electricity into torque. However, a less-explored class of motors produces torque by transferring energy electrostatically by way of contact, corona discharge, and/or induction. Because it is difficult to accumulate large electrostatic charges without breaking down, electrostatic motors have proven to be most effective where small size and high speeds are needed. Electrostatic motors can operate on currents as low as one-billionth of an ampere, and have even been operated directly from atmospheric electricity when an antenna is used.

Because of the extreme simplicity of construction, a quiet dependability of operation can also be expected. Figures 10-1 and 10-2 show examples developed in the 1890s.

One prolific experimenter with electrostatic motors at the turn of the century was Howard B. Dailey. A sample of his work is described in Fig. 10-3. Another type, working from atmospheric electricity, is shown

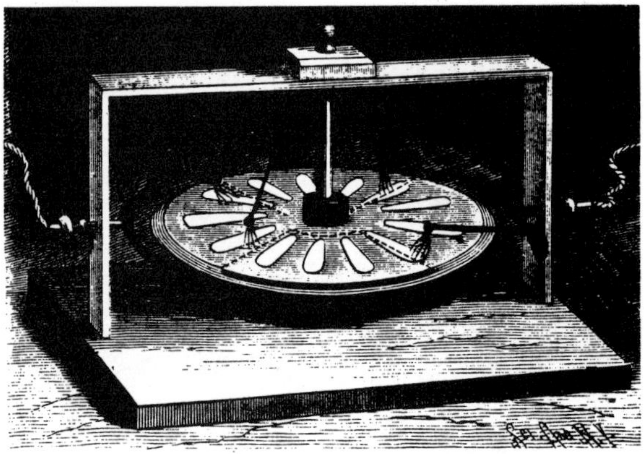

10-1 Disc-type electrostatic motor. *Scientific American*, 1891

10-2 Cylindrical electrostatic motor. *Scientific American*, 1891

in Fig. 10-4. In this case, the antenna needs to be well insulated from moisture because of the small charges involved. In the event of thunderstorms, this same antenna must be grounded well for safety's sake.

Building an electrostatic motor

In my experiments with electrostatic motors, I used a setup like Dailey's, except that my rotors were constructed with thick fiberglass or acrylic discs 2 to 4 inches in diameter.

Coat the rim of the rotor with high-voltage (corona) varnish. When the varnish becomes tacky, roll the disc edge in a pile of #60 grit aluminum-oxide grains (sand blaster's abrasive). These grains make the rotor edges semiconducting, and give a large surface area for charge storage. Instead of thick discs, drums of acrylic can be used. The larger surface area means greater torque for a given diameter.

Static electric top

The continuous action electric top here described furnishes an interesting demonstration of static electric attraction and repulsion.

The top is a disk of stiff mica $4^5/_8$ inches in diameter mounted between two small buttons of wood or vulcanite upon a slender axis made from a piece of steel knitting needle. The pointed lower end projects 1-18 of an inch below the disk and rests in a small indentation worked with the point of a file in the upper end of the vertical glass standard made from a piece of druggist's acid rod. To give the disk a finished appearance, a broad band of red insulating varnish made by dissolving red sealing wax in alcohol is applied to the disk's circumference.

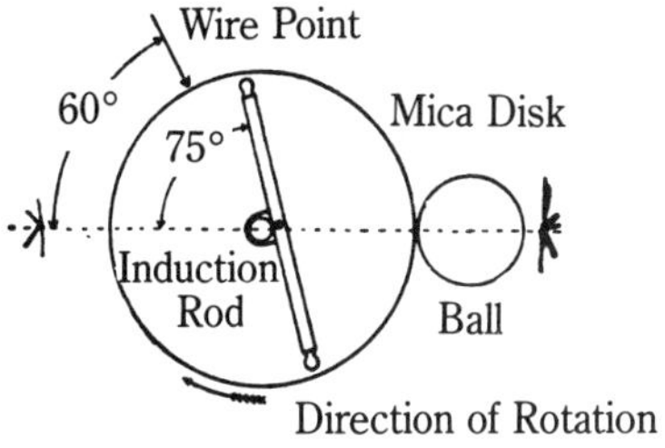

DIAGRAM OF STATIC TOP

At the edge of the disk and as close to it as possible without touching is a $1^1/_2$ inch polished brass or aluminum ball carried by a second support of glass. Rising from the base of the instrument at the left is a curved brass discharge wire arranged so that its pointed upper end approaches the disk's edge very closely from a horizontal distance, at a place about one third the circumference of the disk from the ball.

Beneath the disk, attached by a small brass fixture to the glass stem near its top and on the side nearest the ball, is a polished metallic induction bar with rounded ends, made of $^1/_4$ inch round brass rod. This rod is $4^5/_8$ inches long and is so placed that its upper surface is about $^5/_8$ of an inch below the disk. The angular position of the induction rod and wire discharge point relative to the ball is important. The most effective arrangement is that shown in the diagram, in which the wire point and induction rod make angles respectively of 60 and 75 degrees with an imaginary line drawn through the centers of disk and ball.

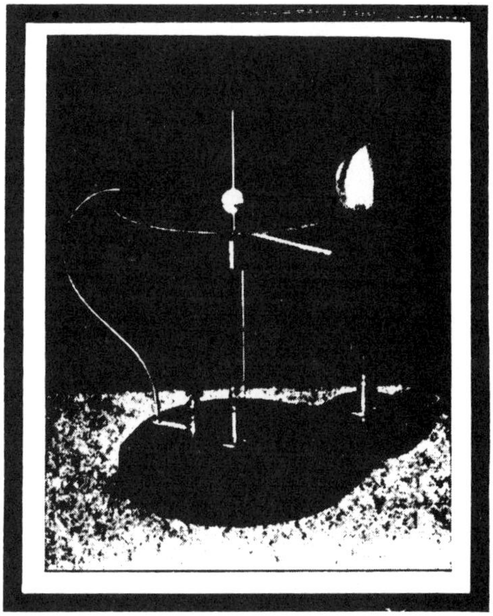

STATIC ELECTRIC TOP

In operating, the insulated ball is connected to the negative pole of a static machine—that pole which gives the brush effect on its collecting combs—the positive pole being joined to the discharge wire through the binding post on the base. The disk, held in position for a few seconds with the fingers placed lightly upon the upper end of its axis, immediately begins a swift rotation, when the finger may be removed and the top will spin at high speed. That position

10-3 Static electric top. *Popular Electricity*, 1912

of the disk at the discharge wire receives along its edges and adjacent surface position electrification, causing repulsion from the point with simultaneous attraction from the negatively excited ball. Rotation ensues, the charged sections of the mica arriving at the ball, giving to it positive electricity and receiving negative in a hissing stream of minute sparks. These parts are impelled forward by the similarly charged ball, until within the attracting influence of the positive wire when the cycle is repeated.

The precise nature of the influence exerted by the brass rod below the disk is somewhat obscure. Through some inductive process its presence seems to effect a certain needed balancing of the acting forces; without it the top indulges in violent gyrations and soon tumbles off.—H.B. DAILEY.

10-3 Continued

Static Motor

Here is a contrivance which will amuse and interest the young electrician. It consists of a small piece of number fourteen gauge copper wire (A), balanced on a pivot (B). Two wires (C) and (D) are supported so that their ends are very close to the ends of wire (A). Connect the lead-in from an aerial to one of these wires, the ground wire to the other, and if everything has been properly made, and if there is a sufficient charge of static on the aerial, the movable wire (A), or armature, will commence to revolve.

For best results (A) should be about $1^{1}/_{2}$ inches long, the distances between the ends of the stationary wires and the ends of (A) being about one thirty-second of an inch. Contributed by S. C. W

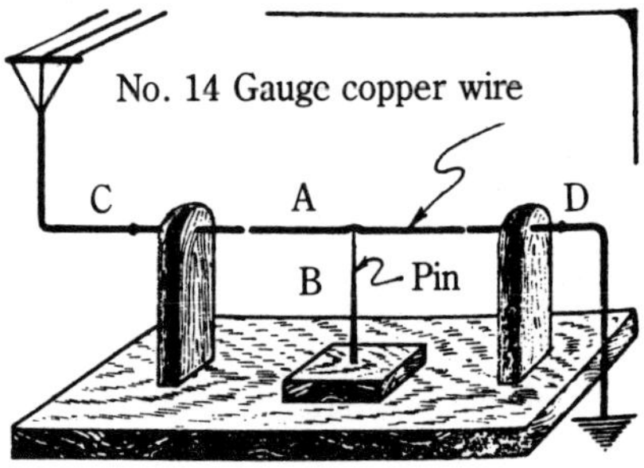

A static motor operated from the ether waves collected by an aerial.

10-4 Atmospheric electric motor. *Practical Electrics*, 1924

Following along this line, I decided to try grinding wheels as rotors, especially since they are balanced at the factory. Aluminum-oxide grinding wheels produce good results. A 1/4-inch-thick by 2-inch-diameter grinding wheel reaches a speed of about 10,000 rpm in less than a minute using a Wimshurst generator as a power supply!

In general, test the semiconducting rotor for its resistance using your homemade electroscope. Hold one side of the rotor (grounding it) and touch the opposite edge to the charged electroscope terminal. The leaf should fall through an angle of 10 degrees during a 60-second interval. This very high resistance ensures that any charge applied to a rotor will distribute slowly over the surface. This disequilibrium in charge produces torque.

If you are researching this subject, study electrode geometry; semiconducting varnishes, waxes, and sulfur coatings; and methods for increasing the surface area of the charged portion of the rotors. Because of their efficiency and compactness, these simple, dependable motors have been used increasingly in the space program.

Caution: Because of the very high speeds involved, surround the rotors with a heavy-walled plastic cylinder. Rotors have been known to explode. This danger is especially present when using the very high voltages produced from the 14-inch Wimshurst machine.

11
Electrohorticulture

ELECTROHORTICULTURE IS THE BRANCH OF HIGH-VOLTAGE RESEARCH THAT deals with the effects of electric fields on growth rate, quantity, and quality of vegetation. Experiments in this field have been traced to 1746 when Mr. Maimbray of Edinburgh, Scotland, studied the growth of myrtle trees. However, serious work on a large scale was forced to wait until the development of high-voltage influence machines, transformers, and better insulators.

In electrohorticulture, also called *electroculture*, electricity is applied to water sprayed on plants, to metal-coated seeds, or by way of air-to-earth electric fields using overhead wires. By far, most of the research concerns overhead wires charged with either high-voltage direct current, high-frequency alternating current (as with the Tesla coil transformer), and low-frequency periodic pulses of direct or alternating current at high potential.

From 1900 to 1915, considerable work was done with the influence machine. One such experimenter was Professor Selim Lemstrom of Helsingfors (at that time in Russia). In 1904, he published a book, *Electricity In Agriculture and Horticulture*, describing his theories, equipment, and results from many years of large-scale experiments. In 1903, he received U.S. patent number 720,711 describing his cylindrical influence machine (see Fig. 11-1 and Table 11-1). This generator, driven by a 1/10 hp motor employed a glass cylinder 30 cm in diameter by 40 cm in length. The generator was quite reliable over extended periods of operation.

11-1 Lemstrom's patented generator. *Electricity in Agriculture and Horticulture*, 1904

Figure 11-2 shows Lemstrom's generator, located in a building for protection from the weather. Porcelain insulators on wood poles support $1^1/_2$-mm galvanized wire running lengthwise and $^1/_2$-mm wires laid crosswise 1 meter apart. The small wires are barbed with points. (See Table 11-1 for the results using this arrangement.) From 1910 to 1929, this research was reported in several scientific journals. The voltages were increased and wire grids were raised 8 to 12 feet above the ground for easier cultivation.

Some of the benefits claimed in these extended tests included yield increases from 21 to 65 percent, increased sugar content in fruits and vegetables, richer colors in flowers, and especially improved resistance to drought and diseases. Even though the current-density requirement is tiny for setting up an effective electrical field (less than $^1/_2$ milliamp per acre or 1.2×10^{-11} amps per square cm), this current is still 50,000 times the natural average air-earth current density!

After 1929, research rapidly declined in the United States and Britain. It is not clear whether the reduced research resulted from the eco-

Table 11-1. Lemstrom's test results for several crops. *Electricity in Agriculture and Horticulture, 1904*

CHEMICAL ANALYSIS OF CROPS AT ATVIDABERG WITH AND WITHOUT ELECTRICAL TREATMENT.

[NOTE.—The increase per cent. of a substance is calculated after the formula $\frac{(a-b)100}{b}$, where *a* represents the quantity which was received under electric current, and *b* the quantity received outside the electrical current.

SUBSTANCE.	RYE.			BARLEY.			OATS.			OATS separated fm. meslin.		
	Under electric current.	Outside electric current.	Incr'se per cent.	Under electric current.	Outside electric current.	Incr'se per cent.	Under electric current.	Outside electric current.	Incr'se per cent.	Under electric current.	Outside electric current.	Inc. or dec. percent
Mineral substance (ash)	2·128	2·440	—	2·979	2·950	—	3·712	3·739	—	3·899	4·107	—
Ether-extract (raw fat)	2·516	1·799	—	2·018	2·443	—	4·831	5·358	—	5·255	5·339	—
Cellulose	2·296	2·379	—	5·605	5·015	—	11·136	9·434	—	9·944	11·655	—
Nitrogen free extract	81·730	83·922	—	77·158	77·502	—	67·781	69·429	—	69·002	66·853	—
Raw proteid matter	11·270	9·460	19·13	12·240	12·090	12·41	12·540	12·040	4·15	11·900	12·040	- 1·16
TOTAL	100	100		100	100		100	100		100	100	
Per cent. of Nitrogen :												
Total nitrogen	1·803	1·513	19·17	1·958	1·935	1·19	2·007	1·927	4·15	1·904	1·927	- 1·19
Amide nitrogen	0·218	0·110	—	0·094	0·061	—	0·093	0·112	—	0·086	0·131	—
Albuminoid nitrogen	*1·427*	*1·249*	14·25	*1·567*	*1·599*	2·00	*1·795*	*1·660*	8·13	*1·724*	*1·661*	3·8
Total digestible nitrogen	1·645	1·359	21·05	1·661	1·660	0·06	1·888	1·772	6·55	1·792	1·792	0·0
Non-digestible nuclein	0·148	0·154	—	0·297	0·285	—	0·119	0·155	—	0·112	0·135	—
Digestibility of nitrogen	91·8	89·8	—	84·8	85·8	—	94·0	91·1	—	94·1	93·0	—
In 100 parts of Nitrogenous Matter:												
Amides	12·7	7·2	—	4·8	3·2	—	4·6	5·8	—	3·6	6·8	—
Digestible protein	79·1	82·6	—	80·0	82·6	—	89·4	86·1	—	90·5	86 2	—
Non-digestible protein	8·2	10·2	—	15·2	14·2	—	6·0	8 1	—	5·9	7·0	—

From this analysis it follows that the electrical air current has produced an increase of the proteid matter in the rye of 19· per cent., in the total quantity of nitrogen 19·2 per cent., in the albumen of 14·3 per cent., and in the digestible nitrogen of 21·1 per cent. In the barley the electric air current only produced in the raw proteid matter an increase of 12·4 per cent., in the total quantity of nitrogen 1·2 per cent. In the oats, on the contrary, it has produced an increase of 4·2 per cent. of proteid matter, of 8·1 per cent. in the albumen, and of 6·6 per cent. in the digestible nitrogen ; whereas in the oats separated from the meslin an increase of albumen of only 3·8 per cent. was produced. All crops under the electrical treatment had been improved in quality to a considerable extent. For rye this improvement can be taken at 20 per cent., for barley at 12 per cent., and for oats at 10 to 12 per eent.

The above analysis was made in the Agriculture Economic Laboratory at Helsingfors (under the direction of Prof. Arthur Rindell) by candidate of philosophy Mrs. Lilly Wendt.

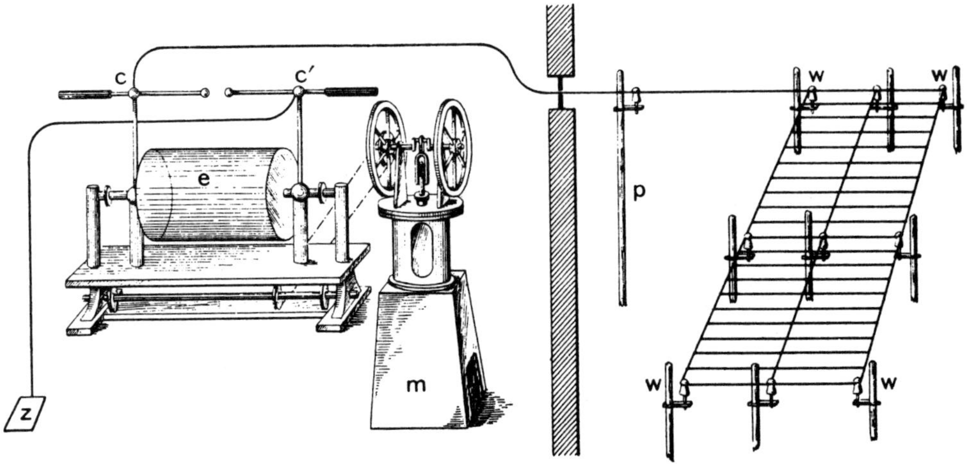

11-2 Charging setup for outdoor plots. *Electricity in Agriculture and Horticulture*, 1904

nomic depression or because of fears about overproduction. A study of the history of this unusual aspect of agriculture would itself be most interesting.

If you are interested in this area of electrification, study the patents issued—for example, U.S. patent number 1,268,949 of June 1918 to Reginald Fessenden and U.S. patent number 1,952,588 of March 1934 to Kenneth Golden.

Our existing "fence-charger" technology, so popular on farms today, might be modified for some electroculture experiments. However, be careful to avoid causing interference to the radio bands—this is an FCC violation that carries stiff fines.

12
Electrotherapeutics

THE BRANCH OF MEDICAL TREATMENT THAT USES ELECTRIC FIELDS FOR diagnosis and cure of disease is called *electrotherapeutics*. Electrotherapeutics involve the study of electrobiology and especially electrophysiology. Just as plants respond to electric fields, so do animals and people.

The first recorded mention of electrical treatment appears during the first century A.D. when the electric torpedo fish produced numbness to the legs and feet, relieving pain from gout. During the 1700s, electrical experimenters carefully studied the torpedo fish, trying to find the source of its electricity. As electrical science developed during the eighteenth century, more experiments were attempted to treat diseases in human subjects—first with static electricity and later with high-frequency currents. In the 1800s, both galvanic batteries and Faraday's transformers were used for specific types of electrotherapy.

The explosive growth of electrical science from the 1890s to the 1920s provided many more instruments and techniques for the electrotherapeutic physician. Both in Europe and North America, science journals dealing with electrotherapy and radiology appeared, and professional associations were formed. Most physicians who used electrotherapeutics were from the regular allopathic school. When drugs failed to help a patient, then electrical methods were employed.

Shown in Fig. 12-1 is a popular electrostatic generator most often used by American physicians. Note the tray containing electrodes and Leyden jars.

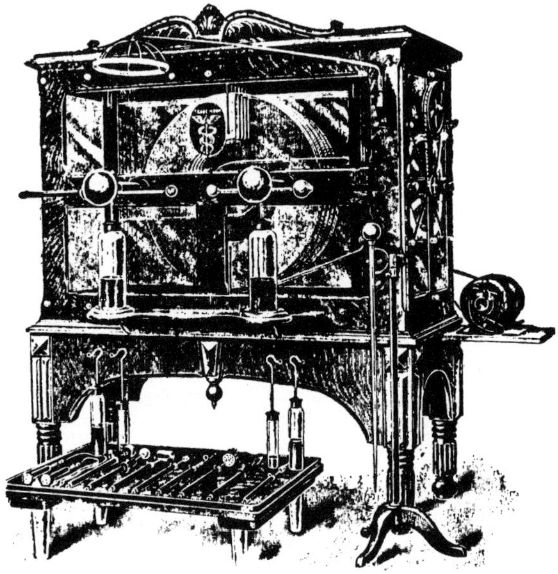

12-1 Electrotherapy generator. *Essentials of Modern Electro-Therapeutics*, Strong, 1908

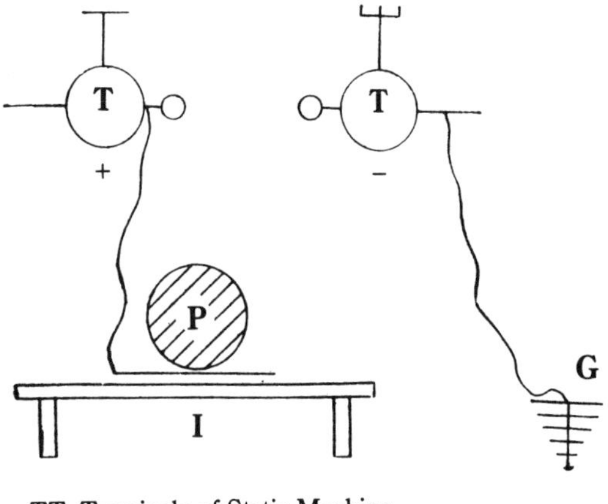

TT, Terminals of Static Machine.
I, Insulating Platform.
P, Patient.
G, Ground Connection.

12-2 Illustration of usage. *Essentials of Modern Electro-Therapeutics*, Strong, 1908

Figure 12-2 shows one of the many setups for using the Holtz or Wimshurst machines. Figure 12-3 shows one of the generator advertisements from the turn of the twentieth century.

During the 1920s, an increasing interest in alternating current as a new technology, especially at radio frequencies, resulted in electrotherapeutic transformers using Tesla and Oudin coils. At this time, several researchers were beginning to study the electrical aspects of physiology.

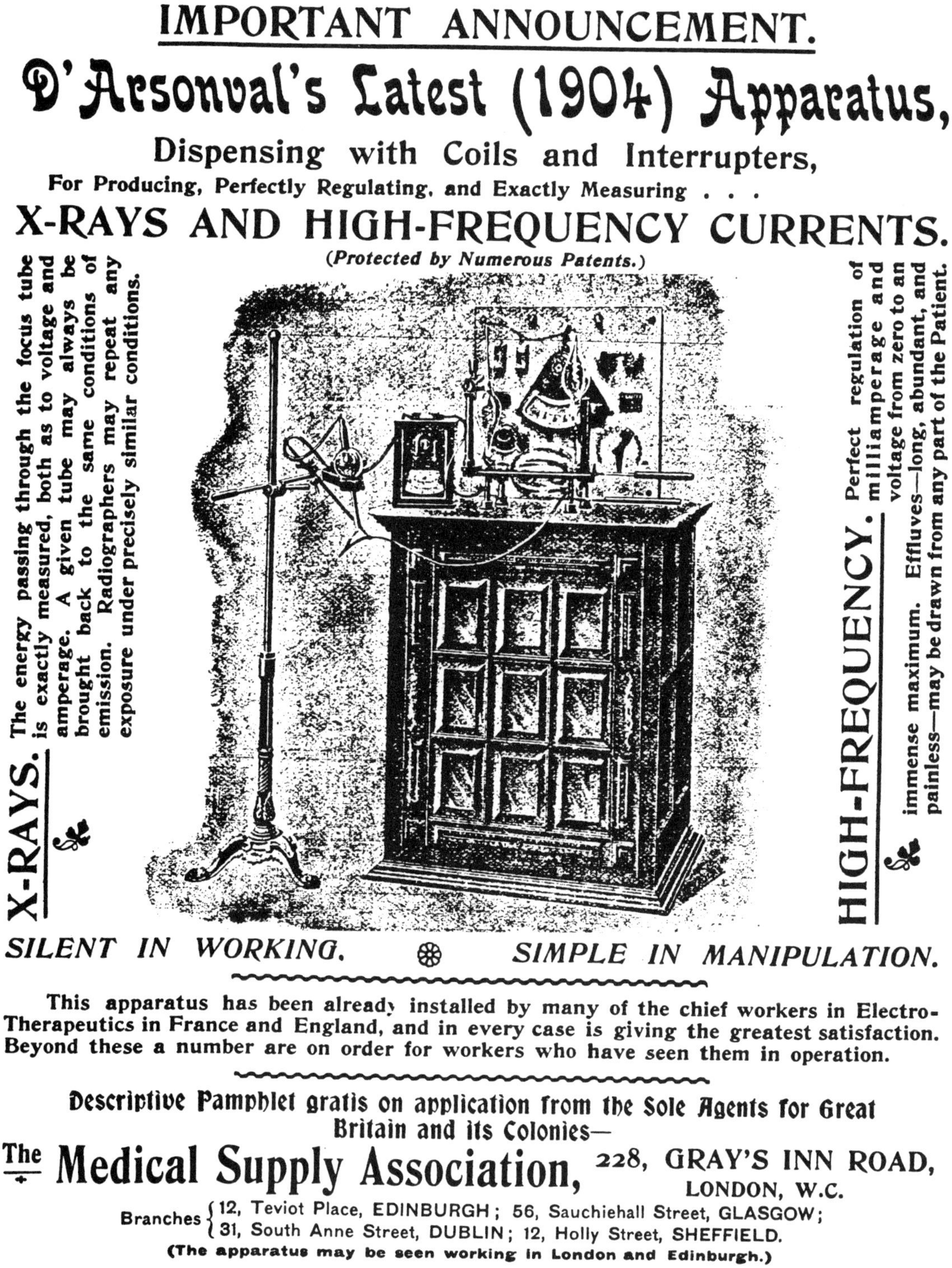

12-3 A typical electrotherapy advertisement. *Journal of the Rontgen Society*, 1905

One result from these experiments was the introduction in 1929 of the electroencephalograph (EEG), which measured the electrical activity of the brain. However, in spite of considerable interest in the electrical aspects of biology, the use of electrotherapy as a recognized branch of medicine declined, especially in America. After some time, even discussing electrobiology with physicians was considered blasphemy. In spite of the well-developed methods, equipment, and efforts expended, the term *electrotherapy* is rarely listed in modern English encyclopedias.

Explaining how electrotherapy disappeared requires a careful study of the powerful political and economic forces that developed during the early 1900s. In addition, excessive claims for electrotherapy during the 1920s brought charges of quackery by the allopathic medical associations. So, the prestige of electrotherapy suffered. Also, the practitioners could not explain in scientifically acceptable terms how it worked. From the 1940s to the present, a physician who used electrotherapy could have his license revoked.

In spite of the general loss of this innovation in medical practice, several vestiges still remain in Western culture. In addition to the development of the EEG is *electrolysis* for the removal of hair, the *pacemaker* for heart patients, *diathermy* and *microwave heating*, *nuclear magnetic resonance*, and last, the most uncivilized of electrotherapeutics, *electroshock therapy*.

In other parts of the world, especially in eastern Europe and Asia, there has been a greater toleration for alternatives to the drugs-as-medicine monopoly that rules in the West. As a result, many of the developments in electrotherapy and diagnosis are fortunately still being used overseas.

It is interesting to note how the Western medical establishment has responded to the Far-Eastern concept of health. For example, Western physicians once denounced the ancient practice of acupuncture as merely having a "placebo effect." However, its use for pain relief during advanced cases of surgery dispelled this superficial explanation.

The issue behind both acupuncture and electrotherapy is the quantification vs. the qualification of health. Orthodox Western medical practice treats the body as a mechanism but neglects studies of energy flow in the body. In recent years, there has been an amalgamation of disciplines that treat energy flow and health. Electroacupuncture, for example, stimulates specific points on the body with electricity to induce changes in bioenergy flow or to remove a blockage to this flow.

Ancient Eastern concepts of health are beginning to meet with Western methods of practice, so we can expect to see a more holistic approach to bodily health, with greater attention paid to the feeling and quality of life, reflected through the use of preventative treatment.

The present crisis in orthodox Western medical practice, involving both chronic and contagious diseases, creates a public demand for access to alternatives. Employing medically unconventional methods should also give rise to a tolerant review of the concepts in electrotherapy, in light of the recent advances in biophysics.

13
High-voltage humans

TO COUNTER THE AUTHORITATIVE CLAIMS DENYING THE IMPORTANT ELECTrical functions in the human body, I provide two cases of anomalous electrophysiology. Through the centuries there have been several eyewitness accounts of this phenomenon. One account was mentioned in the *Annals of Electricity, Magnetism and Chemistry* (vol. II, 1838).

Figure 13-1 describes the case of an ''electrified woman.'' The article finished by saying that the lady was about 30 years old, of a delicate constitution and nervous temperament, had sedentary habits, and seldom had been confined to her bed by sickness.

My second and more recent example (Fig. 13-2) was found in *Science and Invention* (1920).

The Clinton convicts were also mentioned in the *New York Times*, Monday, April 5, 1920, page 1.

Very recently, there have been reports from the *New China News Agency* of high-voltage humans. Dr. Chang Gee Sun of Beijing has been actively studying this area of electrophysiology. Although there might be living examples in North America, the science journals as a rule, ignore such difficult-to-explain anomalies.

If you are interested in the recent innovations in electrobiology, read *The Body Electric* by Robert Becker and Gary Selden (1985). If you wish to research the interesting history of electrotherapy, visit the Bakken Library and Museum in Minneapolis, Minnesota. The electrical exhibits are primarily from the years 1700 to 1900, and include many of the electrostatic generators mentioned in this book.

A lady of great respectability, during the evening of the 25th of January, 1837, the time when the aurora occurred, became suddenly and unconsciously charged with electricity, and she gave the first exhibition of this power in passing her hand over the face of her brother, when, to the astonishment of both, vivid electrical sparks passed to it from the end of each finger.

The fact was immediately mentioned, but the company were so sceptical that each in succession required for conviction, both to see and feel the spark. On entering the room soon afterward, the combined testimony of the company was insufficient to convince me of the fact until a spark, three fourths of an inch long, passed from the lady's knuckle to my nose causing an involuntary recoil. This power continued with augmented force from the 25th of January to the last of February, when it began to decline, and became extinct by the middle of May.

The quantity of electricity manifested during some days was much more than on others, and different hours were often marked by a like variableness; but it is believed, that under favorable circumstances, from the 25th of January to the 1st of the following April, there was no time when the lady was incapable of yielding electrical sparks.

The most prominent circumstances which appeared to add to her electrical power, were an atmosphere of about 80° Fah., moderate exercise, tranquillity of mind, and social enjoyment; these, severally or combined, added to her productive power, while the reverse diminished it precisely in the same ratio. Of these, a high temperature evidently had the greatest effect, while the excitement diminished as the mercury sunk, and disappeared before it reached zero. The lady thinks fear alone would produce the same effect by its check on the vital action.

We had no evidence that the barometrical condition of the atmosphere exerted any influence, and the result was precisely the same whether it were humid or arid.

It is not strange that the lady suffered a severe mental perturbation from the visitation of a power so unexpected and undesired, in addition to the vexation arising from her involuntarily giving sparks to every conducting body, that came within the sphere of her electrical influence; for whatever of the iron stove or its appurtenances, or the metallic utensils of her work box, such as needles, scissors, knife, pencil, &c. &c., she had occasion to lay her hands upon, first received a spark, producing a consequent twinge at the point of contact.

The imperfection of her insulator is to be regretted, as it was only the common Turkey carpet of her parlor, and it could sustain an electrical intensity only equal to giving sparks one and a half inch long; these were, however, amply sufficient to satisfy the the most sceptical observer, of the existence in or about her system, of an active power that furnished an

13-1 A high-voltage human. *Annals of Electricity*, V. 2., 1838

uninterrupted flow of the electrical fluid, of the amount of which, perhaps the reader may obtain a very definite idea by reflecting upon the following experiments. When her finger was brought within one sixteenth of an inch of a metallic body, a spark that was heard, seen, and felt, passed every second. When she was seated with her feet on the stove-hearth (of iron) engaged with her books, with no motion but that of breathing and the turning of leaves, then three or more sparks per minute would pass to the stove, notwithstanding the insulation of her shoes and silk hosiery. Indeed, her easy chair was no protection from these inconveniences, for this subtle agent would often find its way through the stuffing and covering of its arms to its steel frame work. In a few moments she could charge other persons insulated like herself, thus enabling the first individual to pass it on to a second, and the second to a third.

When most favorably circumstanced, four sparks per minute, of one inch and a half, would pass from the end of her finger to a brass ball on the stove; these were quite brilliant, distinctly seen and heard in any part of a large room, and sharply felt when they passed to another person. In order further to test the strength of this measure, it was passed to the balls by four persons forming a line; this, however, evidently diminished its intensity, yet the spark was bright.*

The foregoing experiments, and others of a similar kind, were indefinitely repeated, we safely say hundreds of times, and to those who witnessed the exhibitions they were perfectly satisfactory, as much so as if they had been produced by an electrical machine and the electricity accumulated in a battery.

The lady had no internal evidence of this faculty, a faculty sui generis; it was manifest to her only in the phenomena of its leaving her by sparks, and its dissipation was imperceptible, while walking in her room or seated in a common chair, even after the intensity had previously arrived at the point, of affording one and a half inch sparks.

Neither the lady's hair or silk, so far as was noticed, was ever in a state of divergence; but without doubt this was owing to her dress being thick and heavy, and to her hair having been laid smooth at her toilet and firmly fixed before she appeared upon her insulator.

As this case advanced, and supposing the electricity to have resulted from the friction of her silk, I directed (after a few days) an entire change of my patient's apparel, believing that the substitution of one of cotton, flannel, &c., would relieve her from her electrical inconveniences,* and at the same time a sister, then staying with her, by my request, assumed her dress or a precisely similar one; but in both instances the experiment was an entire failure, for it neither abated the

* It is greatly to be regretted that the spark had not been received into a Leyden bottle until it would accumulate no longer, and then transferred to a line of persons to receive the shock.—Ed.

13-1 Continued

intensity of the electrical excitement in the former instance, or produced it in the latter.

My next conjecture was, that the electricity resulted from the friction of her flannels on the surface, but this suggestion was soon destroyed when at my next visit I found my patient, although in a free perspiration, still highly charged with the electrical excitement. And now if it is difficult to believe that this is a product of the animal system, it is hoped that the sceptics will tell us from whence it came.†

13-1 Continued

One of the First of the Phenomena Noted In the Case of "Botulinus Poisoning," Caused By Eating Decayed Canned Salmon, Was That the Body of the Patient Had Become Highly Electrified. He Was Unable, for Example, to Throw a Piece of Paper In the Waste Basket, the High Electric Charge In His Body Attracting the Paper to His Hand.

AS per schedule, the case of the thirty-four convicts at Clinton Prison, Dannemora, N. Y., who became poisoned by eating canned salmon, and thereafter developt remarkable electrified propensities, was fed to us for several days by the ever-busy newspapers, under the captions of "human magnets" and what not. The facts in the glaring case are here persented for the first time.

The following details relative to *botulinus poisoning,* which took place at this institution, February 20th, 1920, are cited in a letter which we have received from the chief physician at Clinton Prison, Dr. Julius B. Ransom. Dr. Ransom says:

Among Other Things the Electrified Patient Was Able to Move a Suspended Steel Tape Measure and Also to Attract the Filament Of An Incandescent Lamp Towards the Side Of the Globe.

"Dr. Rosneau, of Harvard University, did not make any investigations of the *electric phenomena* and only came into the case with reference to the *botulinus poison,* as it was a rather large group of cases and opportunities for study were unusually good. Of course the newspaper reports were garbled and exaggerated as they usually are when they attempt to report scientific matters. The newspaper accounts were taken from a report made by myself to the Superintendent of Prisons, setting forth the history and development of 34 cases of *botulinus poisoning, due to the eating of canned salmon.*

"During the course of these cases it was discovered by accident that peculiar static electric power had developed in the patients. It was discovered in this manner. One of the patients who was convalescing crumpled up a piece of paper, I imagine in both hands, and attempted to throw it in a waste basket; it absolutely refused to leave his hand. From this time on experiments were made, and the matter was reported to me, and I found that *every* case of *botulinus poisoning* developed this strange power, and that neither the attendants nor nurses associated with them had any such power. All sorts of experiments have been tried and it was found to be a constant condition; that is, that this peculiar power of creating a magnetized (electrified) field by rubbing the hands together, which puts them in circuit, will electrify different objects, so that they will retain that electrification for many hours. For instance forms of paper, such as newspapers, and ordinary correspondence paper when electrified by these patients and thrown against the wall adhered and clung to any object for many hours. By again rubbing the hands together and rubbing the electric light bulb the filament will begin to vibrate very rapidly and follow the motions of the hands, and attach themselves to the side of the bulb with a good deal of sparking at the base of the filament. The compass needle of a surveyor's instrument can be rotated with any piece of paper electrified by these patients. A steel tape suspended, will feel the magnetic field in a remarkable manner and sway from side to side.

"What relation there can be between the botulinus toxin and this phenomena of course is difficult to identify; it has been suggested that it is the *dryness of the skin* which prevents the ordinary passing out or dissipation of the electric currents from the body; *but the patient submerged in bath tub performs the same phenomena as when clothed!* The ability to electrify is proportioned to the severity of the disease; as the patient convalesces he gradually loses this power and when quite well loses it altogether.

Another Phase of the Electrified Paper Phenomenon, Due to the Patient's High Potential Electric Charge Occasioned by Botulinus Poisoning. A Sheet of Paper Electrified by the Patient Would Remain Against the Wall for Hours. He Was Also Able to Move the Compass Needle of a Surveyor's Instrument.

"I might mention further that all these cases were ataxic and developed peculiar reflexes. Many of them were almost entirely blind and had paralysis of the upper lid "Ptosis." Of course, in *botulinus poisoning* the nervous system is about the first to suffer; one thing is quite clear, therefore, static manifestation is closely linked with the disturbance of the central nervous system and represents, no doubt, simply a much higher degree of static storage in the body than is usual."

Electricians Argued That If the Patient Was Placed In a Tub Full of Water, That the Charge Would Disappear, But Strangest of All It Did No Such Thing—and the Patient Was Still Able to Attract a Steel Tape Measure or Other Object By Electro-static Attraction.

13-2 The electric convicts. *Science and Invention,* 1920

14
Cold light

BY *COLD LIGHT*, I MEAN THE PRODUCTION OF LIGHT WITHOUT NECESSARILY the evolution of heat. Work in this field has been inspired by the many examples provided in nature including fireflies, glowworms, auroras, St. Elmo's fire, and earthquake lights.

Since the incandescent lightbulb has an energy efficiency of about 8 percent and the fluorescent bulb an efficiency of 15 to 20 percent, any improvements would, of course, greatly conserve our energy resources (and reduce air conditioning loads). Even though cold light has been closely studied for over 200 years, the subject is one of the least-understood physical phenomena.

The auroral light

The largest and most spectacular display of nature's cold lights are the aurora borealis and aurora australis. The luminous flickering red, yellow, blue, and green lights are seen near the northern and southern polar regions. The relationships between electrical activity and the auroras were most carefully studied near the turn of the twentieth century by Professor Selim Lemstrom of Finland, who in 1883 reported his experimental results (see Fig. 14-1).

Lemstrom succeeded in artificially producing the rare low-level aurora (where streamers reach to the ground) by arranging a ground-insulated flat spiral of wire that covered 900 square meters on top of a hill near Kultala, Finland. A single beam of cold light formed over this installation and extended to a height of 400 feet; the spectroscope showed a greenish-yellow spectrum, with a 5,569-angstrom wavelength of varying intensity. This is the only known experiment that successfully reproduced the properties of the aurora on a large scale.

In an age of exotic, expensive "high" technology, we should marvel at how much Professor Lemstrom accomplished by the simplest of methods, intuitively directed. His experiments show a marvelous, elegant frugality.

On a smaller scale, the Norwegian scientist Kristian Birkeland was able to simulate auroral light. He used a positively electrified copper sphere in a vacuum chamber, with a negatively charged disc at the side of the box emitting charged particles. The sphere represented the earth in space and the disc functioned as the sun.

His main work, *On the Cause of Magnetic Storms* (1908), features photos of his experiments (Figs. 14-2 through 14-4). In Birkeland's experiments, power was supplied at 15,000 Volts dc at 500 milliamps. The sphere diameter varied from 8 to 40 cm, and the chamber was evacuated down to a few hundredths of a millimeter.

A VERTICAL SHEAF OF LIGHT OBSERVED, DURING A DISPLAY OF THE NORTHERN LIGHTS, ABOVE A SYSTEM OF WIRES ON THE TOP OF PIETARINTUNTURI, NEAR KULTALA, FINNISH LAPLAND. (Reproduced from *La Nature*.)

14-1 A vertical sheaf of light observed over Lemstrom's apparatus. *Science* N.S., Vol. 4, 1884.

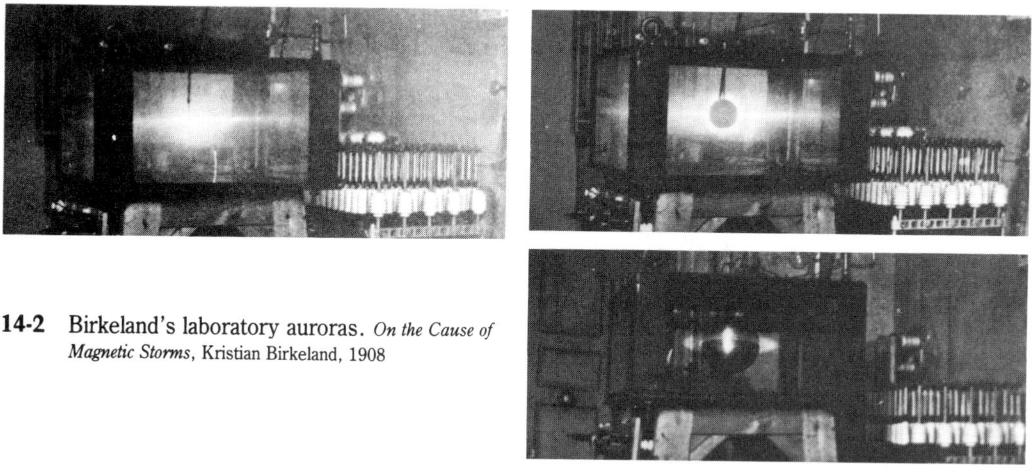

14-2 Birkeland's laboratory auroras. *On the Cause of Magnetic Storms*, Kristian Birkeland, 1908

14-3 Birkeland's laboratory auroras. *On the Cause of Magnetic Storms*, Kristian Birkeland, 1908

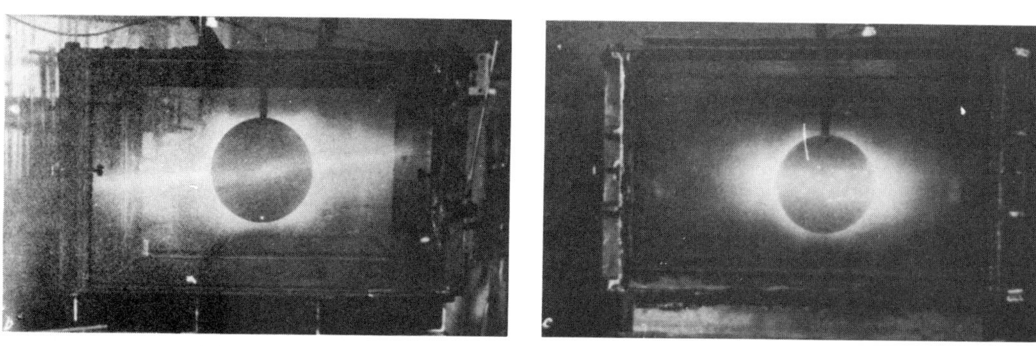

14-4 Birkeland's laboratory auroras. *On the Cause of Magnetic Storms*, Kristian Birkeland, 1908

Earthquake lights

Earthquake lights, which can appear before, during, and after earthquakes, have been recorded since ancient times. The lights take the form of fireballs, streamers, rays of light, flames, or glowing mists of various colors. Several good photographs of this phenomenon have been taken.

The Japanese and Chinese have been careful to preserve good records because the lights can help give short-term predictions for quakes. In one such occasion, on November 2, 1931, at South Hyuga, Japan, the lights appeared on the horizon as divergent rays of blue light. Such a phenomenon can last from seconds to over a minute.

Because of the multiple manifestations of earthquake lights, they cannot all be explained away as power outages or ruptured gas lines. An excellent discussion of these lights is included in *Lightning, Auroras, Nocturnal Lights,and Related Luminous Phenomena* by William R. Corliss (1982).

Possibly related to earthquake lights, but on a smaller scale, is the phenomenon called *triboluminescence*, that is, light produced by impact, friction, or rubbing. In addition to smoky, amethyst, rose quartz, and flint, sugar crystals show brief flashes of light when they are struck or fractured.

A simple, but "illuminating," experiment follows: Allow your eyes to become adjusted to a completely dark room for about 20 minutes. Now chew Wint-O-Green® or Pep-O-Mint® Lifesavers® mints and have a friend observe (or with a mirror observe) the bluish flashes of light in your mouth as you break up the candy. Loaf sugar and rock candy also produce this effect.

Professor Linda Sweeting, from the chemistry department at Towson State University in Towson, Maryland, has been studying the spectrum of "mouth lightning." Sweeting relates the color to the nitrogen in the air and the separation of electric charges by fracturing.

By extrapolation, you can visualize how, on a large scale, triboluminescent rocks such as quartz look when they are fractured or rubbed together during a quake or in a rock avalanche. The tons of material and surface area involved should produce quite a light display. Released flammable gases, such as methane, would produce jets and sheets of flame, quite distinct from cold light.

This branch of cold light, if seriously studied, could help with short-term earthquake prediction and also with locating minerals in the earth's crust by spectral analysis. If a large number of people in earthquake-prone areas were educated on this subject, then it might be possible to classify these light forms.

Invisible phosphorescence

Invisible phosphorescence is the least-studied form of cold light. This cold light was discovered in 1899 by the incomparable Dr. Gustave Le Bon and described in his work, *The Evolution of Forces* (1908). When phosphorescent compounds such as calcium sulphide are exposed once to light then kept in the dark, the materials continue to radiate, sometimes for months. Although these radiations are not visible to the eye, they can be recorded on film. A related form of light is shown in Fig. 14-5.

Figure 14-5 is a slight variation of mine in which the 14-inch Wimshurst machine has both a large 2-inch-diameter wood ball terminal, which is painted with luminous paint, and an uncoated 3/4-inch steel ball terminal.

When you take a photo, you can only see several sparks over the 6-second interval. Imagine my surprise after developing the film to note the continuous bright glow. Evidently streams of paint particles are dislodged by the discharge, but they radiate light, mostly at a wavelength only captured on film (see Fig. 14-6). Le Bon was even able to take photographs through opaque bodies using these invisible radiations!

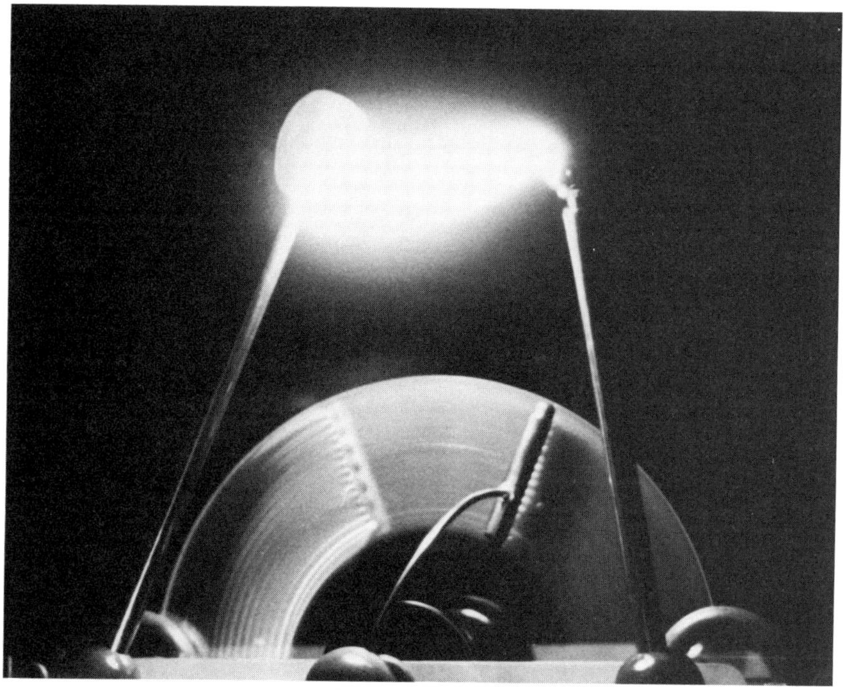

14-5 Invisible electro-phosphorescence. Film: Black & White ASA 400, F/ 1.7, 6 seconds.

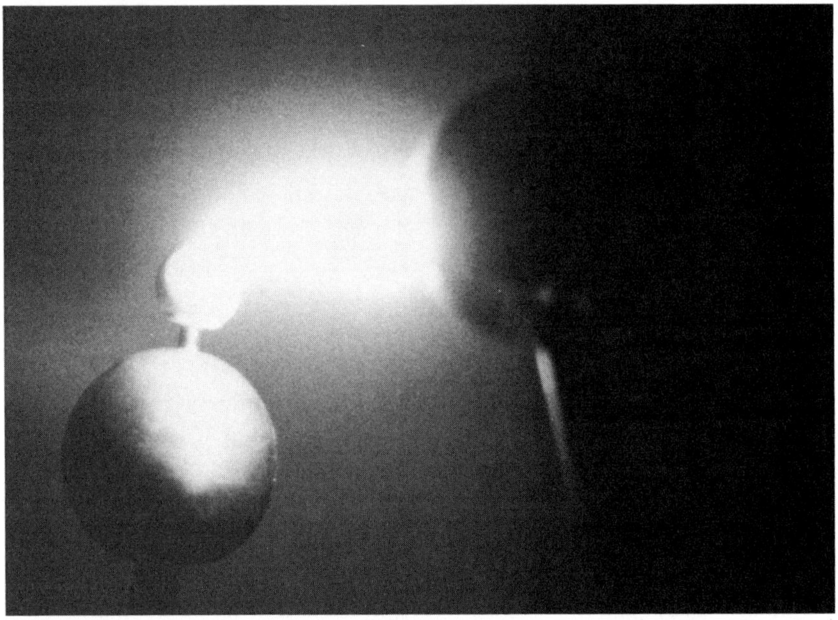

14-6 Invisible electro-fluorescence from a sulfur-coated metal sphere Film: Black and White ASA 400, F/1.7, 6 seconds.

It would be most profitable to continue the research work of a number of pioneers. Those who were heavily involved in the study of phosphorescence and fluorescence include Edmond Becquerel (late 1850s), William Crookes (1880s), Gustave Le Bon (1890s), Johann Puluj (1890s), and Hermann Ebert (1890s).

Building the electric egg

Experimentally, the nature of cold light in controlled atmospheres, such as rarefied gases, might be studied by constructing the "electric egg" (see Fig. 14-7).

The egg-shaped glass chamber has two metal rods. Terminate each rod with a ball, which can be brought close or separated as desired. Remove the egg from its stand and join it to a vacuum pump. Control the pump's flow with the stop cock at the base of the chamber.

At a vacuum pressure of 60 mm, electric discharges appear, as shown in Fig. 14-7. At approximately 3 mm of mercury pressure, the discharge appears as a red-tinted luminous sheaf issuing from the positively charged ball. The negatively charged ball and rod are enveloped by a layer of bluish-purple light. A water aspirator or a discarded refrigerator compressor can provide an adequate vacuum pressure for many cold light experiments.

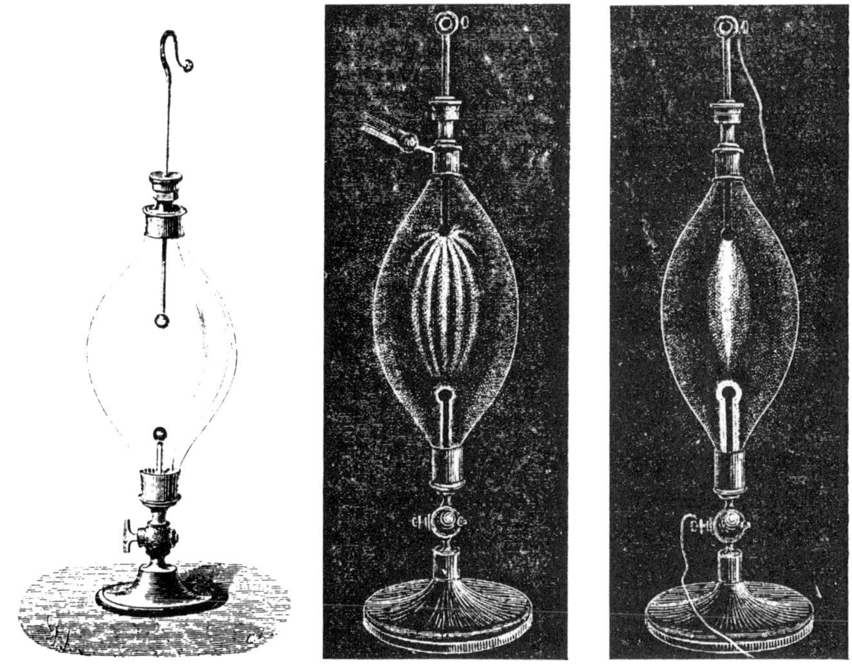

14-7 The "electric egg." *Electricity and Magnetism*, A Guillemin, 1891

Phosphorescent lamps

During 1896 and 1897, several articles appeared in science journals on the phosphorescent lamp developed by Austrian researcher Johann Puluj. His lamp (Fig. 14-8) was shaped like an Edison incandescent lamp, but the wires extended through the bulb, rather than through the socket. Both wires were made from aluminum. The negative pole (cathode) ended in an aluminum reflector-shaped disc. Hanging from the apex of the globe was a small square sheet of mica. The mica surface facing the reflector was painted with calcium sulphide (key ingredient in luminous paint). Radiant electricity converged from the disc onto the painted mica anode. The anode glowed with a brilliant phosphorescent (cold) light.

Caution: When working with vacuum bulbs do not use a very high voltage or a very low vacuum pressure. If either condition occurs, X-rays, which result from the sudden stoppage of high-speed electrons, will be produced. Moderate pressures and voltages will keep the lamps cool and reduce deterioration of the phosphorescent compounds.

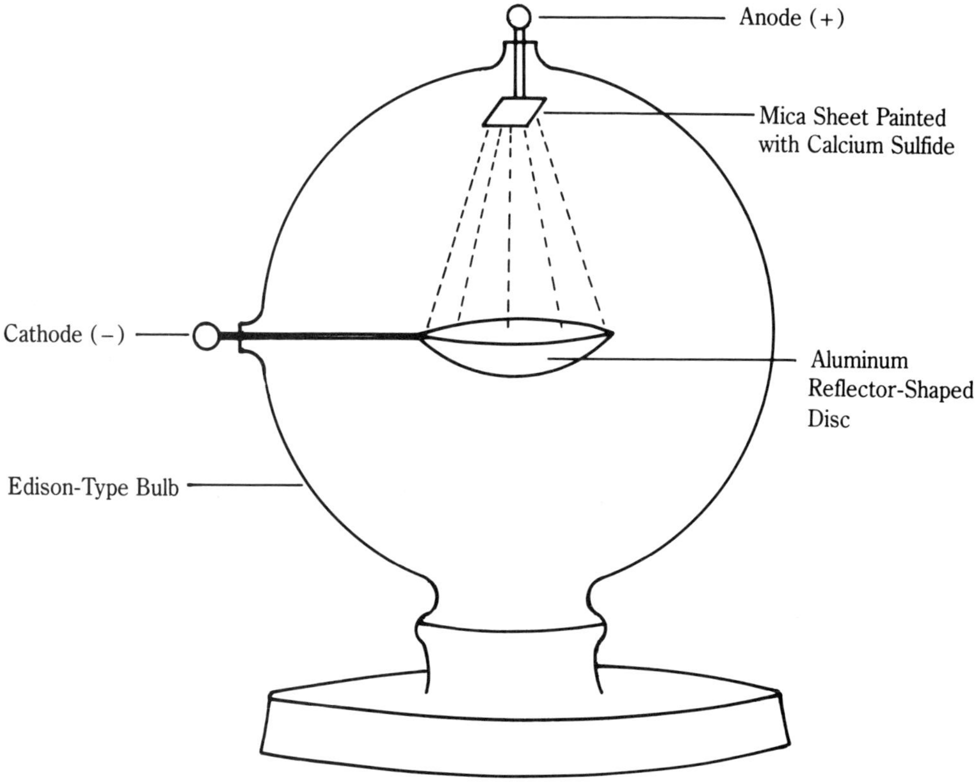

14-8 Puluj's electro-phosphorescent lamp.

15 Miscellaneous experiments

A hair-raising experiment

IN FIG. 15-1, OUR RADIANT BEAUTY, DIANA, IS CHARGED TO ABOUT (+) 300,000 volts. She holds a 1/4-inch-diameter aluminum rod rounded at both ends and brings one end near the positive terminal, while the negative terminal is grounded with a jumper wire. Your subject's hair must be clean and dry. No hair sprays should be used, they can be flammable.

Warning! Leyden jars must be removed for safety. The platform on which our model stands is a thick wood slab with legs made of PVC plastic pipe 2 1/2 inches diameter by 4 inches long. This insulated platform is placed on a large rubber mat to prevent sparkovers. With this arrangement, the insulated person will reach the potential of the terminal.

At this energy level, all exposed human hair stands at attention, and the sensation feels "prickly." A higher potential, 500,000 volts, for example, at the same current, would cause an uncomfortable stinging sensation.

Extrapolating from this experiment, consider the experiences of a bird flying from earth to a high-voltage power line. Its body potential will start at "neutral ground," but as it nears the wire, its electrical potential rises. When it touches the wire, it is at the wire's potential, normally 3,000 to 7,000 Volts (alternating or direct current doesn't affect the result). The bird feels nothing at contact. The electrified wire modifies the properties of the surrounding space.

Besides this hair-raising experiment, you can map out electric fields around the generator by blowing bubbles from on the insulated stand. Bubbles usually fly to the walls or ceiling along peculiar paths. Figure 15-2 illustrates this experiment.

15-1 Author's version of a room-temperature super conductor.

The levitating rocket

The next novel experiment involves levitating lightweight objects in space with electric forces. A number of patents have been granted for scientific toys that use this principle. A charging wand, operated by fric-

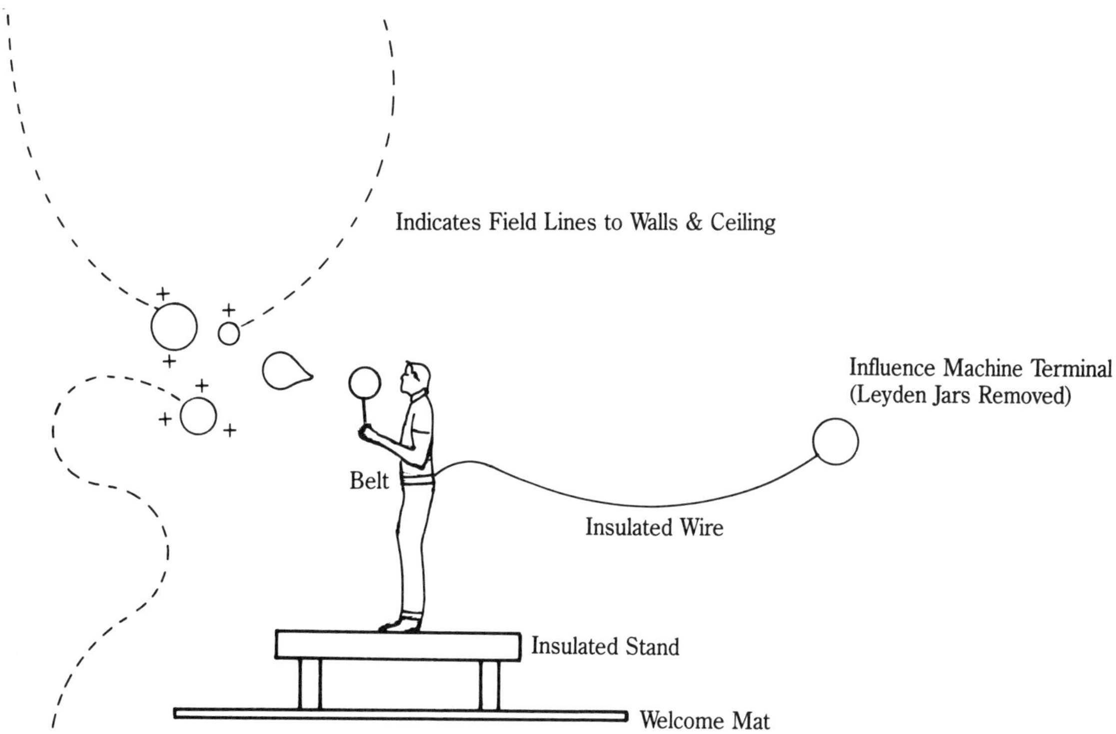

15-2 Mapping electric fields with charged bubbles.

tion, normally provides the electric charge. If you like to make toys that illustrate scientific principles, study U.S. patents 2,018,585 October 1935; 3,497,994 March 1970; and 4,109,413 August 1978. A more "advanced level of play" makes use of the influence machine. The writings of Dr. Gustave Le Bon feature his description of the electrified "rocket."

Figure 15-3 shows the rocket levitating in space between a positive-charged metal sphere and a grounded, pointed metal probe. A painless way to launch the rocket is to charge the end of a 1/2-inch-diameter PVC pipe by touching it against the charged sphere (Leyden jars removed). Now place Le Bon's rocket (Fig. 15-4), cut from very thin aluminum foil, crosswise near the end of the charged pipe. Place the rocket at the mid-point between the sphere and the probe and slowly roll the pipe. This action dislodges the rocket, and it will normally remain suspended in space as if it was held by invisible springs.

The modified levitating rocket

In another experiment, give the rocket a one-quarter twist at the mid-point of its length. When suspended, it will spin like a top. Next, slowly

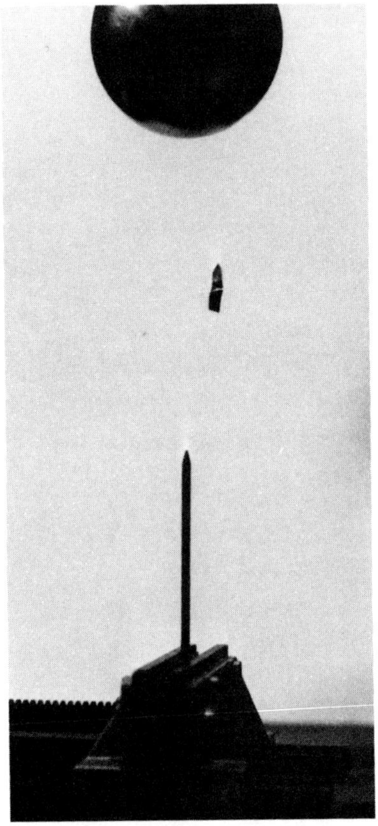

15-3 Le Bon's electrified rocket suspended in space.

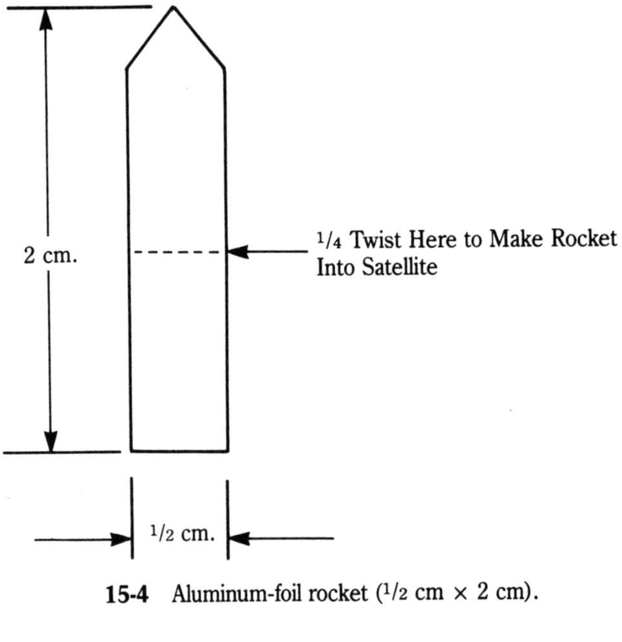

15-4 Aluminum-foil rocket (1/2 cm × 2 cm).

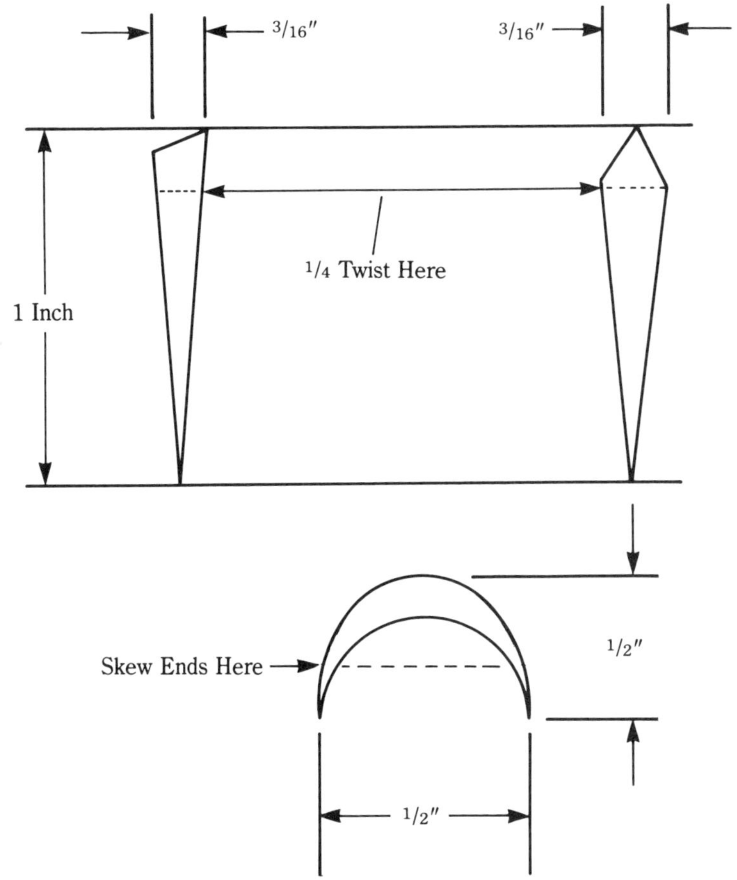

15-5 Various shapes of satellites.

pull the metal probe away. The rocket is now a satellite, as it goes into orbit around the positive sphere. Occasionally, two satellites of the same shape, but of smaller size, can orbit at the same time.

Figure 15-5 shows other shapes that have worked. Besides aluminum foil, try gold leaf, polystyrene foam plastic, and feathers.

16 Electroaerodynamics

IN ELECTROAERODYNAMICS, ELECTRIC CHARGES ARE APPLIED TO HIGH-SPEED vehicles, such as supersonic aircraft, for the purpose of reducing air drag or eliminating sonic booms. In practice, high-speed ions are projected forward from the leading edges of the craft. This corona discharge propagates "upstream," and repels motionless air molecules away from the aircraft's oncoming surfaces.

In effect, this system should keep the relative motion between the vehicle and the adjacent air molecules so low that a shock wave cannot be mechanically produced. Of course, electroaerodynamics, by reducing frictional drag, could greatly conserve fuel resources and reduce air pollution.

In the early 1960s, Horace C. Dudley received U.S. patent number 3,095,167 on his means for increasing the altitude of his electrified model rockets. However, his results need to be conclusively verified by other model-rocket enthusiasts.

In March 1968, an article was written on the experiments of Dr. Gustav Andrew and Maurice Cahn, who worked at Northrop's Norair Division in Hawthorne, California. In a 5-inch supersonic wind tunnel, they were able to propagate a corona glow 6 to 8 inches upstream from a 1/2-inch-diameter charged probe. This experiment proved that a flow of ions could charge the air molecules that precede a supersonic aircraft. If you are interested in attempting this experiment, you should insulate the throat of the wind tunnel well so that the charge is not lost to the walls. In addition to the high voltage, a substantial current will be needed to affect a large volume of air molecules. The design of the electrodes and of the leading edges is a crucial aspect because it facilitates uniform spraying of ions in a forward direction. An efficient influence generator would be quite adequate for use in preliminary small-scale wind-tunnel tests.

Above all, do not be dissuaded from experimenting. A mathematical disproof that indicates the power requirements are excessive for practical purposes cannot cover all design variations. Hence, actual wind tunnel tests are necessary. The improvements made in computer modeling over the last 20 years should facilitate in the design of electrified leading edges. (See *Product Engineering*, Vol. 39, March 11, 1968, pp. 35–36 for a related article.)

17
Countergravitation

AS STATED BEFORE, ELECTRIFICATION APPEARS TO NOT ONLY PENETRATE inner atomic structures, but also alter the properties of space itself. Scientists have long speculated on the relationships among gravity, space, and electricity, but little progress has been made with finding a common denominator among these properties.

Based on my research, the problem hasn't been clarified for two reasons: Scientists have failed to use visual modeling—specifically, how gravity and electrification act; and researchers have misconceptions about the properties of space itself. Physics textbooks often describe gravity as a force that reaches across empty space and pulls two bodies together (i.e. "action-at-a-distance").

This notion has been wrongly attributed to Newton's views, described in *Principia* (1687). On the contrary, Isaac Newton did not, to his credit, reject the existence of an ethereal medium in space. He wrote, "I have no regard in this place to a medium, if any such there is, that freely pervades the interstices between the parts of the bodies." However, since the character of space was not resolved by Newton or his students, the mystical notion of action-at-a-distance gradually developed. Newton himself labeled this concept "absurd."

The ether, whether thought of as tiny subatomic particles or waves, was considered necessary to account for the propagation of forces. However, by the turn of the twentieth century, when the ether theory was most fully developed, the seemingly paradoxical properties of space still could not be resolved. The result was that abstract mathematics began to replace visual modeling. Even though the ether concepts disappeared by the 1930s, since the 1950s especially following the work of Paul Dirac, the interest in the properties of space has revived slowly. However, *ether* has been replaced with such terms as *neutrino flux*, *gravitons* (a quantum unit of gravitation), *soft particles*, *virtual particles*, and *zero-point energy*.

Kinetic gravitational theory

Newton's concept might be considered a static description of gravity. A dynamic or kinetic theory would explain gravity as originating, not in the bodies, but in the space itself! One of the most interesting explanations was that of Georges Louis Le Sage in 1749. His theory might be visualized as follows: Imagine that all space is filled with tiny particles (his term was *ultramundane corpuscles*) traveling in all directions at high speed. Because of their subatomic size, the tiny particles essentially pass through all material bodies. A single body, such as a planet in space, might be pictured as in Fig. 17-1.

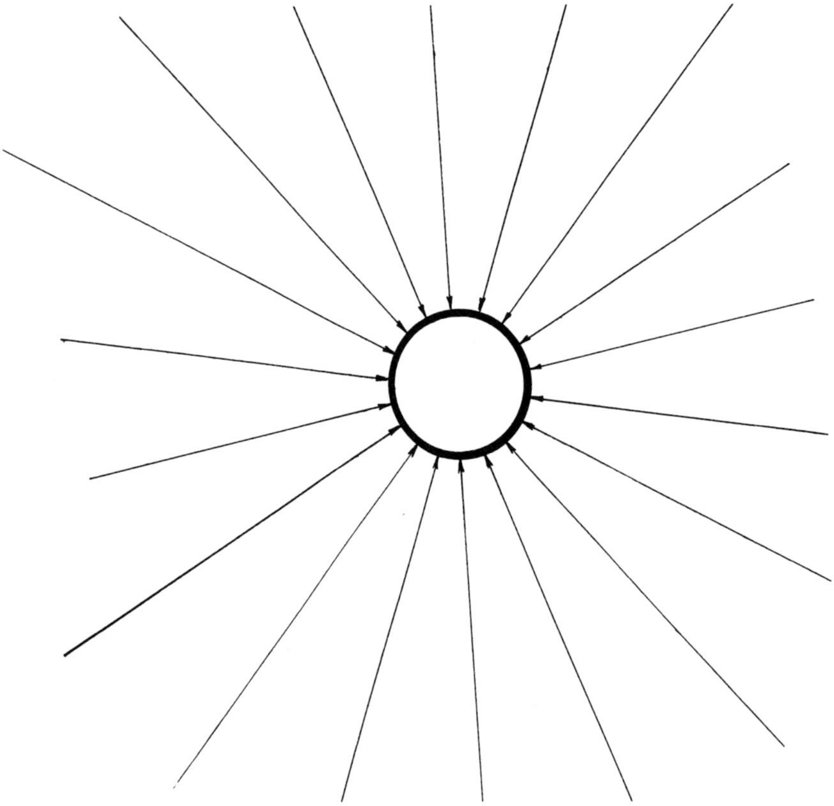

17-1 Action of kinetic space on a singular body.

In the simplified drawing (See also Fig. 17-2), two bodies are pushed together by the superior energy density on the surfaces not facing each other. Each body casts an "energy shadow" on its neighbor so that the ether density is very slightly reduced between the two bodies.

I hope you can now see the importance of visualization as a useful tool in understanding phenomena in physics. These mental models will also help to direct your experimental tests.

You might wonder if there are any natural phenomena or experiments that might shed light on the properties of space. Some optical phenomena, because of the propagation of light, depend on the electrical and magnetic properties of space. Perhaps halos and coronas, sometimes seen around the sun and moon, are indicators. Many anomalous light forms have not been explained.

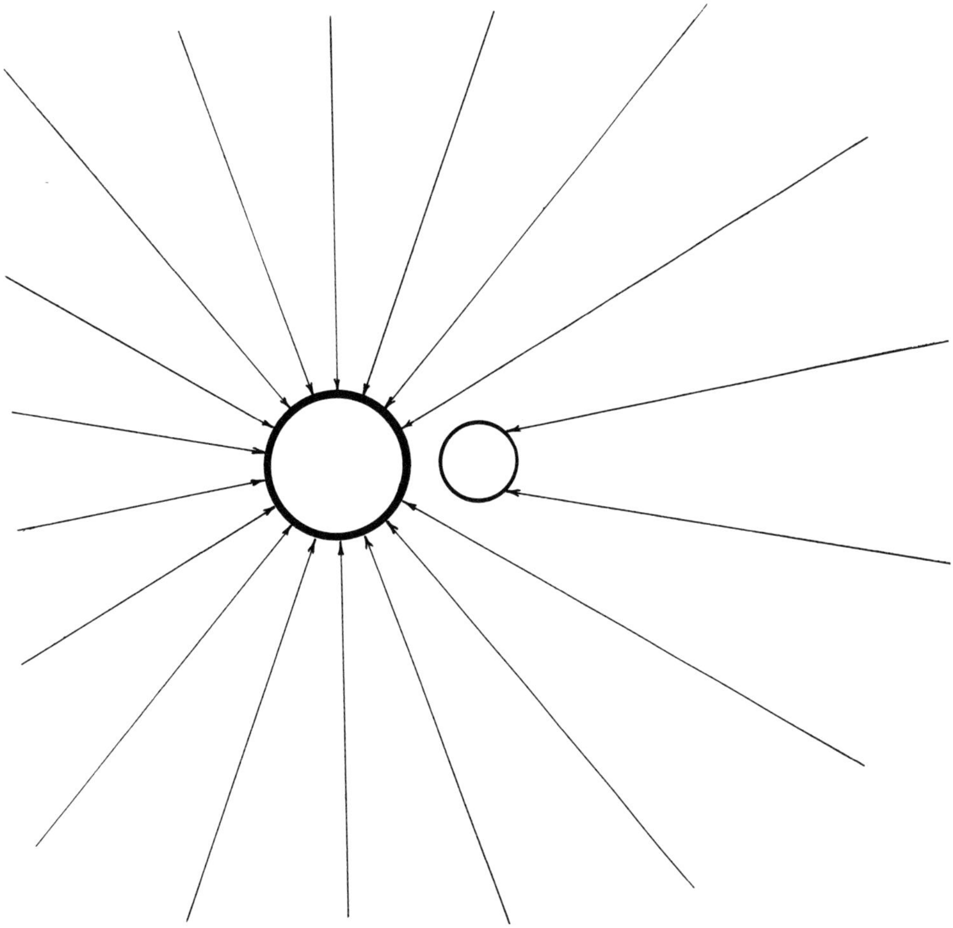

17-2 Gravity—a push phenomenon.

As far as experiments in this subject matter are concerned, American industrialist, Charles F. Brush, published the results of his tests from 1914 to 1929. The results showed that rocks composed of complex silicates of protooxides of nickel and cobalt show a spontaneous rise in temperature in the ambient air during calorimeter tests (1927). In other tests, he found that certain metals and compounds can fall at a slower rate (1 part in 140,000) in a gravitational field! Specifically, bismuth and barium aluminates produced the best results (1924). He attributed these strange results to a slight interaction between atomic structures and gravity waves.

Figure 17-3 gives two popular accounts concerning the relationship between electrification and gravitational action.

Dr. Nipher's deflection experiments

This experiment—performed by Dr. Francis Nipher, professor of physics at Washington University, St. Louis, Missouri—is a modification of the Cavendish experiment of 1798. In this earlier experiment, Henry Cavendish used a delicate torsion balance to determine the density of the earth.

The first phase of Dr. Nipher's experiment, performed in 1916 and 1917, is shown in Fig. 17-3. The room had a concrete floor and granite walls, and the equipment was mounted on a massive bench. Thermometers nearby indicated that parts of the apparatus did not vary in temperature from each other by more than 1.5°C. The 1-inch lead ball was suspended with an untwisted silk thread approximately 180 cm. long, and centered inside a 5-inch-square iron box or Faraday Shield. A horizontal slit in the box's side, covered with a glass plate, permitted Dr. Nipher to observe scale deflections with a telescope.

Next to this iron box, he placed an insulated 10-inch-diameter lead sphere, with a copper wire keeping this sphere and the metal box at the same potential. To eliminate errors caused by temperature differences, Nipher used cardboard heat shields and kept the observer's body below and away from the apparatus.

Figure 17-3A shows the normal attraction between the uncharged masses. In Fig. 17-3B an influence generator located in the next room, is joined to the large mass. After about twenty minutes, the 1-inch lead ball slowly moved to the opposite side with a deflection about twice the normal gravitational attraction regardless of the polarity used.

In the last phase of this experiment in 1917, a torsion balance with two large spheres and two small balls gave the same results. Next, the large lead spheres were replaced with charged metal boxes containing cotton batting. This gave no deflection, eliminating electrostatic force as the cause. Finally, the influence generator was replaced

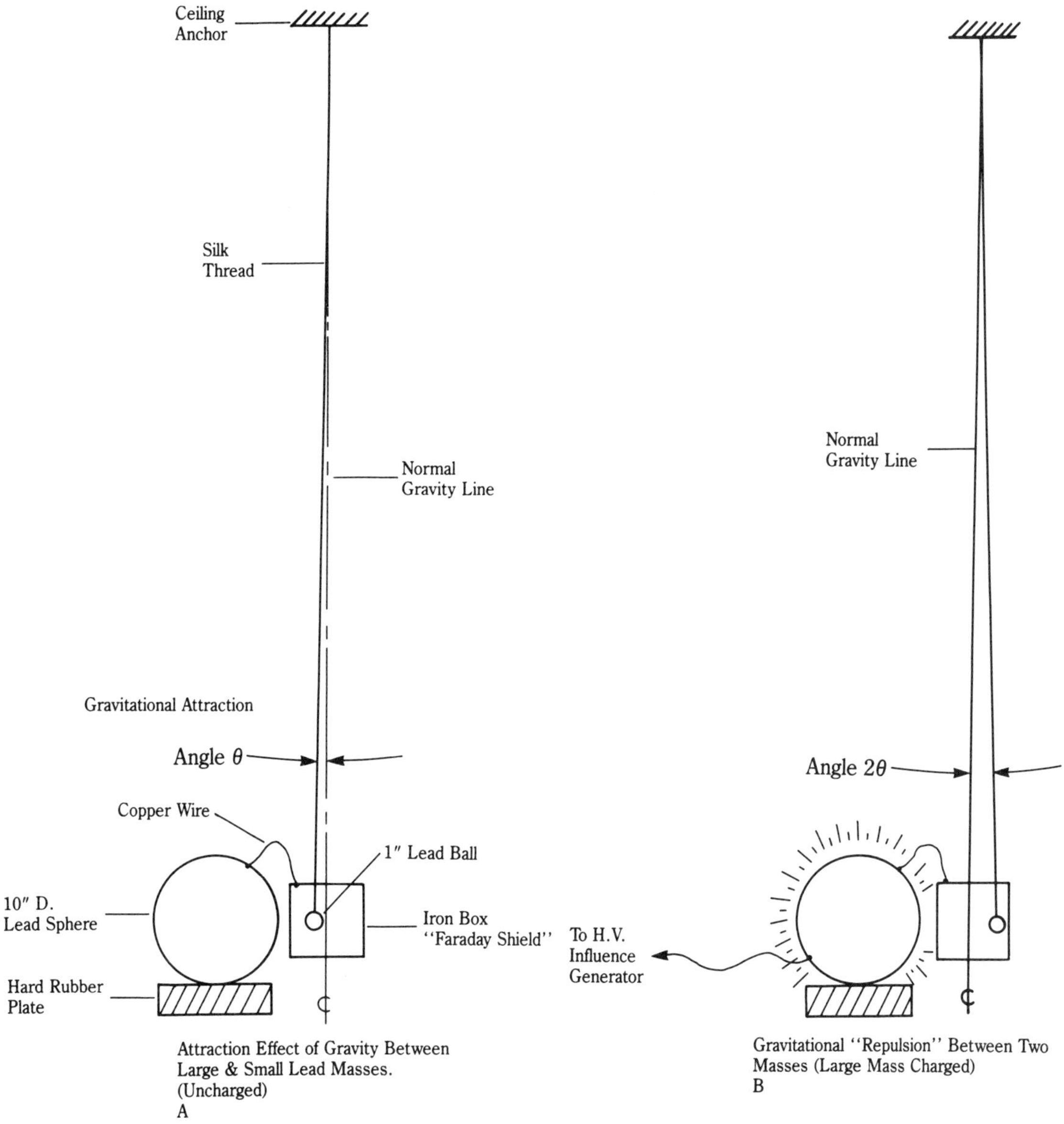

17-3 Dr. Nipher's electro-gravity experiment. *Electrical Experimenter*, March, 1918

with a low-voltage ac current passing through the two large lead spheres. This also produced a repulsion effect, but one of smaller value.

The full details of the experiment were given in *Transactions of The Academy of Science of St. Louis* Vol. 23, 1916 and 1917. Also see *The Electrical Experimenter*, March 1918, for a related article.

Although Nipher's experiments met a deafening silence when they appeared in the scientific journals, no one came forward with an alternative explanation. Also, Nipher was well respected by his colleagues and appreciated for meticulous accuracy in his experiments. The best description of this work is found in *Transactions of the Academy of Science of St. Louis* (1916).

The article from the *Electrical Experimenter* in Fig. 17-4 shows a more popular line of thought.

Overcoming' Gravitation
By George S. Piggott

For some time past there has been quite a controversy going on regarding the subject of interplanetary communication by means of electric waves. I have been very much interested in the above on account of experiments which I have made and data collected pertaining to gravitation effects on high frequency oscillations and electronic discharges in general. A series of experiments which I conducted during the year 1904, caused me to formulate the theory that interplanetary transmission of electrical impulses was an impossibility on account of the sun's resisting and absorbing influence which vertually isolates our planet from all other electrical vibrations of a lesser tension or power.

Gravitation Suspended in Experiments.

The above theorem was arrived at after I had succeeded in sustaining a metallic object in space by means of a counter-gravitational effect produced thru the action of an electric field upon the above object. A strong electric field was produced by means of a special form of generator and when the metallic object was held within its influence it drew up to approximately a distance of 1 mm. from the center of the field, then was repelled backward toward an earthed contact, going within 10 cm. of the same when it was again attracted toward the field's center but this time getting no nearer than 5 cm. from the polar nucleus. This backward and forward movement continued for some time until the metallic object at last came to a comparatively stable position, about 25 cm. from the field's center where it remained until the power was shut off. While the metallic object was suspended, I was able to study the effect of the surrounding field and found by means of a powerful microscope, assisted by the insertion of a vacuum tube within the field, that the metallic object (having of course a certain electrical capacity) became fully charged and gave off a part of said charge to and against the surrounding field which tended to hold said object in space, apparently without any other sustaining influence. Around the outside of the metallic object and extending to a distance of about 1/2 cm. was a completely dark belt or space in which there appeared to be no electrical agitation due, possibly, to neutralization caused by the contact of the large incoming energy supply from the field's center with the small oscillating radiations from metallic object. The ever changing action of attraction and repulsion resulted in the overcoming of gravitation. Going farther I will state that the dark belt above mentioned after many tests gave no sign of electrification, a most astonishing phenomenon, inasmuch as its width was but 1/2 cm. In fact, a dark line was shown in the vacuum tube when it was introduced between metallic object and center of field. It is my firm conviction that somewhere on the outer confines of our planet there exists a similar counteracting belt thru which naught but the gravitational vibrations of the sun penetrate, and these vibrations absolutely annihilate or absorb all other less powerful ones.

Therefore, after making many experiments to ascertain as nearly as possible the absolute facts and conditions as they exist, I have come to the conclusion that all electrical disturbances not due to our own radio oscillations, on this globe are due to the sun's electrical activities in semi-inductional contact with our polar extremities.

17-4 Electric levitation of metallic spheres. *Electrical Experimenter*, July 1920

Details of "Defying Gravity."

The illustrations 1 to 4 will possibly give a fair idea of the apparatus used, and the manner in which the experiments were carried on.

Fig. 1 shows general scheme of arrangement of devices. In the lower left hand corner is shown the "ground contact," which can be turned around and placed in any position found necessary, in fact, when metallic object is in suspension, this *ground* can be entirely eliminated.

I have found that any substance within the limits of my experiments can be held in suspension, viz: water globules, metallic objects, and insulators being among those tried. Some materials such as cork and wood exhibit peculiar activities when suspended; a piece of green maple would not rest in one position within the field, but oscillated backward and forward, continuously, going to the field's center, then back to ground.

Heated materials exhibited equally peculiar characteristics: A silver ball 11 mm. in diameter when heated, remained farther away from the field's center than when at normal temperature; upon cooling it gradually drew up to the position it would occupy if unheated.

Fig. 2 shows a generator of the Wimshurst type (improved), the generating or collecting units being entirely enclosed in an insulating case and operated under a pressure of 3 atmospheres; completely dry air *only*, entering case thru drying device attached to air pump shown in Fig. 1. Interior parts of generator will retain quite a powerful charge for a long period of time.

Fig. 3 illustrates suspension stand and field producing electrode, the latter can be revolved in any direction by means of a spring motor shown on upper section of stand.

The small apertures seen in electrode, which is hollow, are there for the purpose of ascertaining the action of the reduced field

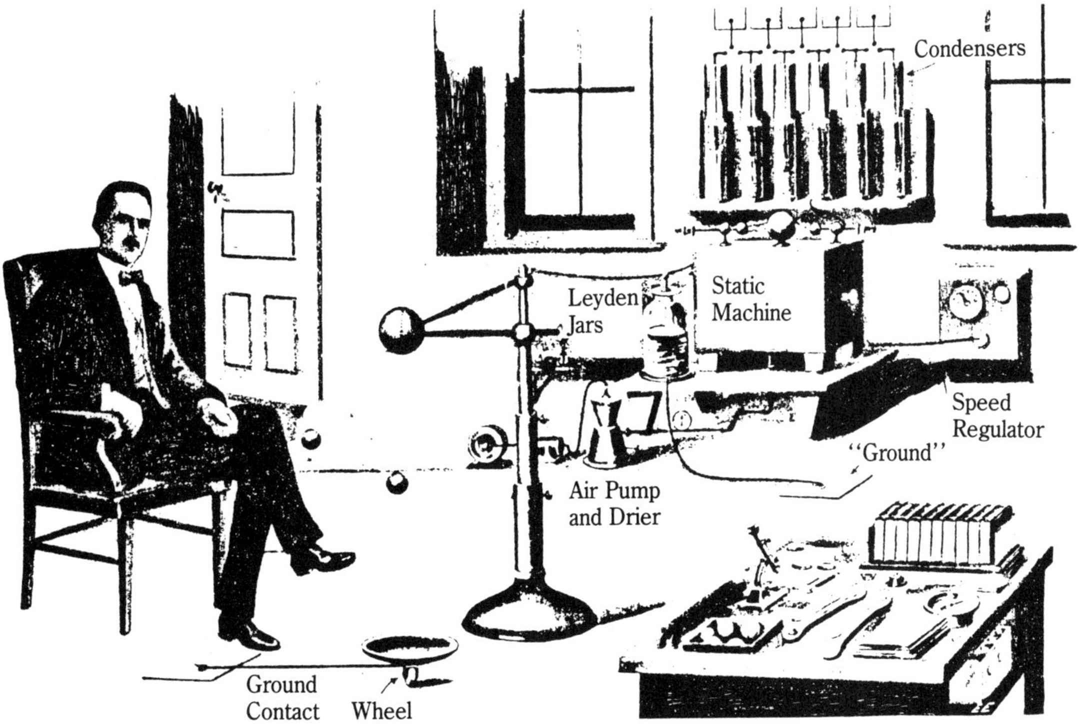

Fig. 1.—This picture shows Mr. George S. Piggott, the author, and his laboratory with the powerful electrical apparatus used, whereby he was enabled to carry on successful experiments in nullifying the effects of gravitation. In other words, he was able to suspend small balls and other objects in the manner shown, the silver balls actually used having weighed 1.3 grams. The diameter of the balls was 11 mm.

17-4 Continued

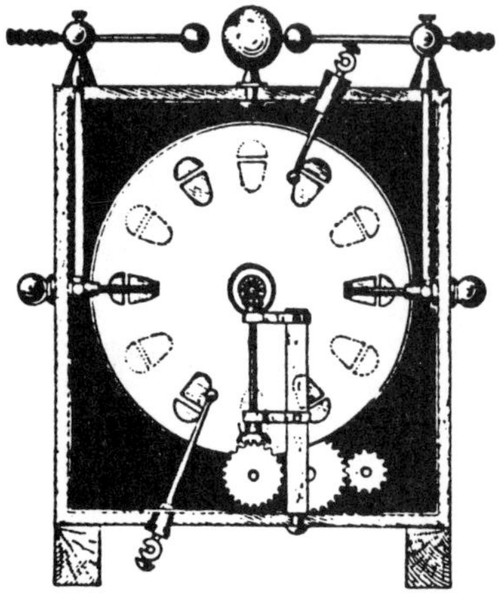

Fig. 2.—Special electro-static machine used by Mr. Piggott in his gravitation nullifying experiments. Which was enclosed in a heavy-airtight compartment, so that it could be operated under several atmospheres of air pressure.

tension at these points, and are also made use of to hold different sized metallic discs, which are cemented to insulating plates, forming condensers, the function of which is to create weak opposite polarities at these points and thus show a reaction on the suspended object and also a greater ocular effect in the vacuum tube.

Fig. 4 is a detailed drawing of the vacuum tube principally used; this is of the spectrum type, without sealed-in electrodes and when introduced into the electric field, glows very brightly at its extremities, especially giving a sharp line bordering the dark space around the metallic object. A very high vacuum is sustained in the tube and it is found necessary to build it of a very perfect insulating glass; the bulb musts be kept absolutely dry on its outer surface.

Different tubes have been used beside the above; corrugated spherical, cone shaped, and cylindrical, with various results.

The electric field produced for suspension experiments is very powerful and

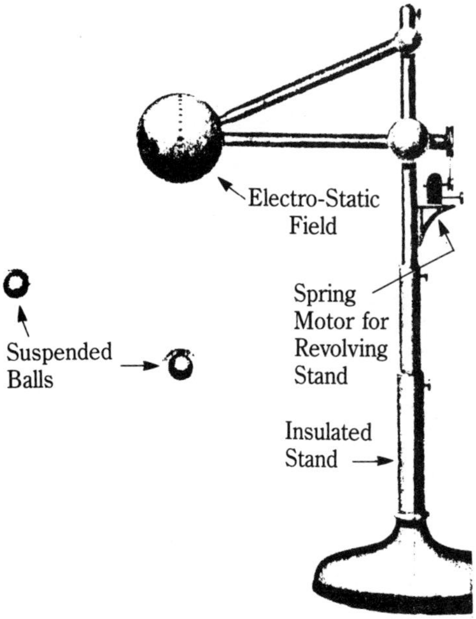

Fig. 3.—A close-up view of the charged metal sphere mounted on a pedestal together with a spring driving motor, whereby the electrode or charged ball could be rotated. The two smaller silver balls are shown as suspended in mid air, the Earth's gravitational pull having been nullified.

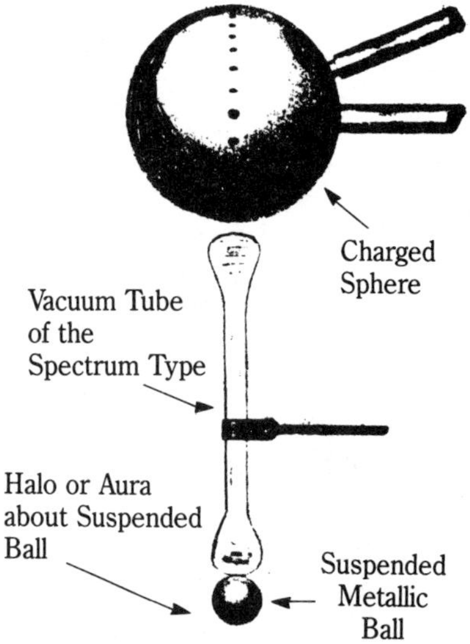

Fig. 4.—Close-up view of vacuum tube of the spectrum type used in studying the aura surrounding the suspended silver balls, while they remained suspended in space.

17-4 Continued

intense, being detectable with a vacuum tube at a distance of over 6 meters (19.68 feet).

In conjunction with the above and drawing an analogy between the same, I am of the opinion that cometary motion is undoubtedly due to the activity of its compositional elements and their susceptibility to changes of polarity, which, when the comet is far distant from the sun, would be opposite in sign to that of the latter, or when in close proximity to the central orb, would be of the same sign and therefore repelled.

All bodies in process of formation possibly have their cometary stage, and doubtless future experiments will reveal this fact.

Actual Results Achieved by Mr. Piggott

The total power required to operate generator, which was run by electric motor, was about 1/4 K. W. generator; the machine voltage was in the neighborhood of 500,000 when the electrodes were separated beyond sparking distance. The electrostatic charge left on the suspension electrode retained the average object in space for a short length of time, about 1 1/4 seconds after machine ceased rotating.

Some objects such as copper and silver balls, which are of course good electrical conductors, and very nearly homogeneous, when falling toward the earth, after power had been shut off, seemed to slow down when they neared same, and hovered about 2 c.m. above contact for approximately 1 sec. of time before striking same; this was due no doubt to the inductional change of polarity which was imparted to balls almost at the instant of earth contact.

The aura, shown in figure 3, near suspended balls (which in this experiment were made of silver) extended outward to a distance of about 1 c.m. and covered about one-half of the upper hemisphere and a trifle more of the lower hemisphere.

This bluish emanation appeared to be made up of numerous infinitesimal dots or darting particles, each apparently separated from the other by *a very narrow, glowless belt.* Everything was, however, in a constant state of agitation and it was quite impossible to get an absolutely perfect view microscopically, of an individual particle. Different substances have different aura both in length and breadth, and also in luminosity.

The silver balls used in these experiments had an actual gravitational weight of 1 3/10 grams (nearly .05 oz., avoirdupois) and were the heaviest objects suspended at this time, their diameter being 11 mm. as before mentioned in another part of this article.

The largest object suspended was a cork cylinder 10 c.m. long by 4 c.m. diameter (approximately 4 by 1 9/16 inches) which had a copper wire pusht thru its center, and extending beyond its ends to a distance of 3 mm. The weight of above cylinder was 3/4 gram (.002645 oz. avoirdupois).

The behavior of metal spheres used in above experiments was a most interesting spectacle, silver and copper balls floated very steadily on one position and when suspending electrode was revolved, would follow and turn slightly axially, but would not revolve entirely around same, there being a peculiar "slipping" effect not entirely accounted for.

17-4 Continued

In Piggott's experiments, the simultaneous appearance of strange luminous halos occurs in conjunction with the effects of levitation. Note that a high-voltage threshhold of about 500,000 volts must be reached before the effect is produced. Of course, by running his Wimshurst in a chamber containing a compressed gas, such as dry air or carbon dioxide, the current output was increased considerably. Oddly, the spheres float. If the phenomenon was simply an electrostatic force, an electrostatic field would first attract, and then repel a metal sphere.

Three other scientists who devoted their lives to kinetic gravitational research are:

- Thomas T. Brown, who extended Nipher's work to include the spontaneous self-motion of capacitors (1929)
- Thomas Jefferson See, who developed the mathematical concepts supporting his Wave Theory of Gravity (approximately 1920 to 1950)
- William J. Hooper, who invented two artificial-gravity field generators using the B × V field (approximately 1968)

18
Unusual electric discharges

THIS CHAPTER FEATURES A FEW OF THE STRANGE ANOMALIES IN electrical science that will be of considerable interest to those making electrical investigations.

Lightning shadowgraphs

Lightning shadowgraphs might be called nature's method of photography. These shadowgraphs are of shadow-pictures of objects projected on nearby surfaces from bright flashes of lightning. For example, the *English Mechanic* (1892) reported an incident in Errol, England. A telephone repairman, while fixing a fused telephone wire damaged by lightning, found an image of the roof of a nearby house on one of the porcelain insulators. Apparently, the brilliant light and vaporization had flashed the image onto the smooth surface of the insulator.

The second example (Fig. 18-1), comes from *Scientific American Supplement* (1904).

Occasionally, these images have been fairly permanent. The image of a lady's face at her bedroom window (watching a thunderstorm) was said to be flashed onto the window pane. After many years, the impression gradually faded away, perhaps by erosion of the glass surface. The chemical condition of the nearby surface is important.

Tornadoes as electrical machines

Several years ago, I experimented with the effects of high-voltage direct-current discharges onto moist semiconducting surfaces. The

Remarkable and rare effects of lightning.

An excerpt from the Annals of the German Hydrographic Bureau furnishes us with a bit of information at once interesting and astonishing in its effects. While on a voyage recently from Hamburg to St. Thomas the second officer of the Hamburg-American liner "Galicia," being on the bridge during a terrific electrical display, observed the following phenomena, which he carefully noted, and which it is our privilege to present to our readers. In advance it may be remarked that all the wood and iron work about the bridge had been painted gray. In changing his position he casually removed his hand from a cabinet on the bridge immediately after a particularly brilliant flash of lightning, and what was his astonishment to notice an exact counterpoint of it in silhouette upon the cabinet, and to add to his amazement the picture remained imprinted fully five minutes. Such a spectacle was well calculated to incite the officer to further observations, which he carried out with like results. Among others he placed an observation instrument upon the cabinet, and waiting his opportunity removed it just after a vivid flash, to find the shadowgraph perfect in detail, even to the cross-hairs over the objective plainly visible upon the surface.

Since the ship's deck was also painted gray, he determined to try a further experiment; and with this in view threw down upon the deck an annular cork life preserver, allowing it to remain untouched for several successive flashes.

In throwing it down, whether with intent or otherwise, the ship's name and hailing port, "Galicia, Hamburg," painted upon the cork ring, fell downward next the deck. When the ring was removed, the shadowgraph was plainly seen, and what was more, the inverted letters in more somber tones could be distinctly read. Until it had entirely disappeared watch told off seven minutes, the additional duration resulting from the effect of the several consecutive flashes. Keenly awakened by a spirit of investigation, the officer experimented upon the galvanized ironwork sustaining the bridge, which was, as before said, also painted gray.

From this he failed to elicit any response, while all the woodwork seemed particularly sensitive. Moreover, it was discovered that success depended upon the wet or moist condition of the painted surfaces; upon dry objects of the same color no pictures were obtained.

In discussing the phenomenon, the annals remark that should an attempt be made to explain the pictures by declaring that the lightning of itself had nothing whatever to do with their appearance, but rather that the different objects placed upon the cabinet and the deck had absorbed the moisture, and caused a dry spot surrounded by a wetted surface, making it distinguishable from its surrounding by a shade of color, it would hardly explain the presence of the cross-hairs over the objective, which are contained entirely within the instrument. Nor would such a hypothesis hold when compared with the experiments upon the ironwork.

A more plausible elucidation of the occurrence would derive from a chemical examination of the constituents of the paint used, which might disclose some phosphorescent properties of the ingredients. Upon request the Hamburg-American Line furnished the German Marine Observatory with some of the liquid, which by some inadvertence or carelessness has been lost before it could be used. Having aroused an interest in the proper accounting for such amazing displays, the government desires that observations be continued, and in cases of recurrences, either some of the paint or some object covered with it, which has given the abnormal results, be sent to the Lighthouse Board for official investigation.

18-1 Remarkable and rare effects of lightning. *Scientific American* supplement, 1904 substances—including granite, marble, agate, limestone, sandstone, white chalk, plaster of paris, slate and unglazed clay—were chosen for their fine porosity and ability to absorb moisture.

A most unusual discharge presented itself while I was working with unglazed clay flower pots. Figures 18-2 and 18-3 illustrate the setup for producing miniature electric tornadoes with a white fireball tip. A 1-inch-long section of plain steel piano wire, 0.015 inch in diameter, is positioned 0.06 inch from the clay surface. The pot sits in a shallow metal pan with water covering the bottom. The power supply is a full-wave

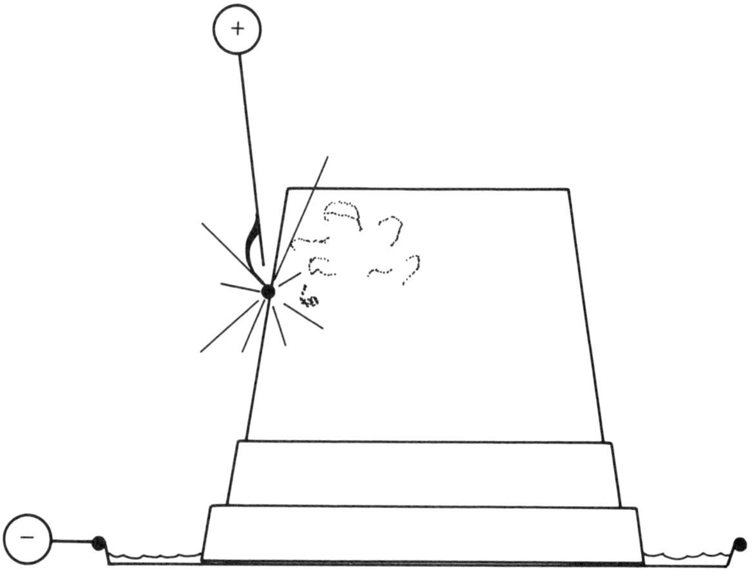

18-2 Miniature electric tornado.

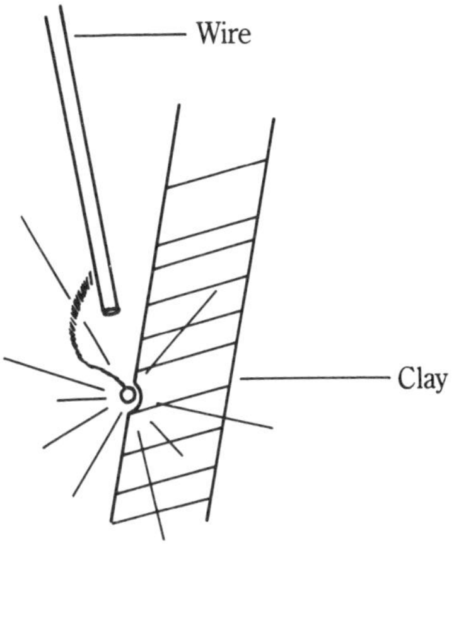

18-3 Miniature electric tornado.

transformer rectifier that has an output of 10,000 Volts dc at about 0.7 to 1.0 milliamp (7 to 10 watts). The wire is made positive, and the pan is negative.

The unglazed clay's properties are crucial. At the Ceramics Research Laboratory, University of Illinois, analysis comparing clay pots indicates that the acceptable clay, on which the vortex forms, has a dry surface resistance of infinity and, when dipped in water and the excess wiped off, a surface resistance of 300,000 ohms. Place the probes of the ohm-meter 1 cm apart for this indication. The clay color is light red; that means it contains a smaller percentage of iron oxide. Thanks to Dr. Relva Buchanan for determining the characteristics of the clay samples!

Moisten the selected flowerpot with your finger dipped in water and place it as shown. Turn on the power and the discharges will remove moisture, increasing the clay's resistance. In the dark, when the fireball forms, the discharge either squeaks or is silent. Hold the probes near the clay and a dazzling pure white ball, about 1 mm in diameter, will form on the pot in an electric tornado-vortex, varying between 1/8 and 3/8 inch long .The ball will slowly traverse the surface in a sinuous movement, seeking out a path of preferred resistance.

When I examined the ball through a #4 gas welder's filter, rays from the fireball were still visible. The amazing thing to me is that the heat from the tiny fireball was so great that it permanently etched a black path into the clay. Figures 18-2, 18-4, and 18-5 show the characteristic track signatures. When the polarity is reversed, the tip of the steel wire (−) often glows white hot, and the clay remains cool.

How does this unusual discharge relate to real tornadoes and waterspouts? Many good descriptions of tornado lights with fireballs and internal lightning bolts exist, but I found only one case with a fireball maintained at the terminal end (as in our experimental condition). One rare account in Fig. 18-6 is from the British journal, *Weather* (1949).

Caution to experimenters using high-wattage power supplies: the evolution of heat might be so great as to cause flash-steam explosions at the clay's surface. Some shielding for your eyes should be included in your setup. Also, be sure that the wire electrode does not vibrate; this action destroys the vortex formation.

If you are interested in tornadoes, their formation, and prediction, consider the following avenues and questions:

- Does a relationship exist between tornado "alleys," and ore deposits and underground streams?
- Based on our experiments, electric polarity is important. Does the anvil cloud, therefore, have a high-positive charge where the

vortex forms and a negative-charge concentration on the ground below the cloud?

- Would studying the miniature electric tornadoes with high-speed infrared and ultraviolet photographic methods, and the doppler shift (for determining the rate of fireball rotation) help in understanding the mechanics of vortices?
- Is there a connection between trailer parks and tornado formations? Trailer roofs present a large aluminum surface to the sky; this might strengthen the air-earth electric field.

Indispensable to any study of tornado anomalies is the excellent source book, *Tornadoes, Dark Days, Anomalous Precipitation and Related Weather Phenomena* (1983) by William R. Corliss.

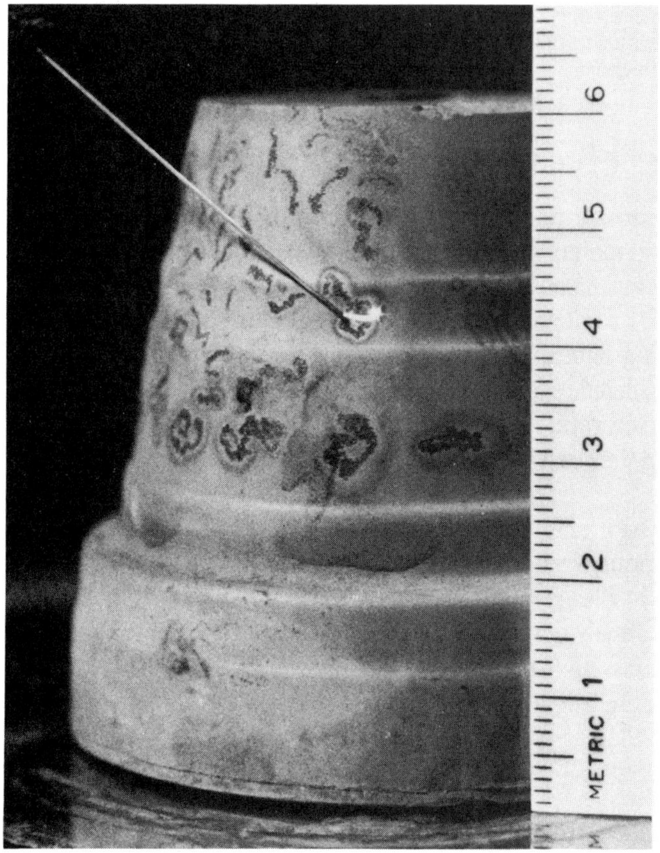

18-4 Fire ball-tipped tornado on moist clay. Film: ASA 400, f/5.6, 1/15 second, scale in millimeters.

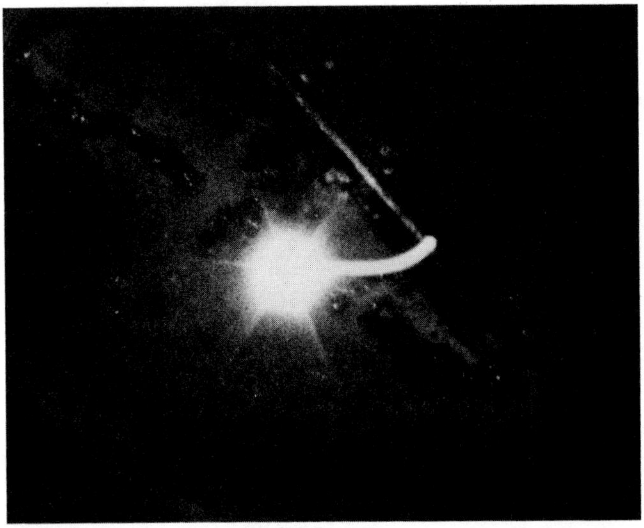

18-5 An enlarged view. Camera: Pentax SLR and Vivitar close-up lenses #4 and #2, with ultraviolet filter. Focal distance 7 inches; power required: 10,000 vdc at 0.7 milliamps.

The electrical entities

The various names for self-sustaining electrical forms have included electric meteors, bolides, fireballs, and lightning balls. I prefer the term *entities* because they are self-sustaining embodiments of electrical force.

In its ordinary form, an electrical entity appears as a smooth sphere emitting light in often beautiful colors. However, its bewildering variations include spheres that are smokey black globes with spikes or diverging rays, double and triple balls connected by luminous threads, and globes with long tails, rod-shaped lightning bolts, and luminous rings.

The "bible" of eyewitness accounts is, at present, the 170-page document, *Der Kugelblitz*, by Walther Brand (1923) NASA translation (1971). Please see the bibliography for other sources, since Brand's work mainly covers the classic ball shape.

Drawing from several sources, which include both naturally formed and artificially formed electric entities, I have sorted out several conditions that are most often present in the formation stages. The list includes:

- The presence of suspended moisture, such as a fine mist. The air is usually at the saturation point. Also, the presence of excessive dust, soot, or finely divided metals and semiconducting stone particles is frequently noted.

- The presence of hygroscopic porous or fibrous bodies including adobe, stone, wrought iron, or metal oxide on conducting wires, plaster and chalk, talc, gypsum, or clay.
- Discontinuities in metal conductors, including rapid changes in cross section, sharp bends, projecting points, and in the case of chain conductors, poor contact between successive links.
- A very slowly rising electric-field intensity that reaches a high potential without causing an actual breakdown. It evidently takes some time for the air molecules surrounding the site to become mobilized to an electrified state. Suspended drops of moisture might help this process to occur.
- Occasionally, the initial entity formation requires a shock in the form of a thunderclap or falling object. Perhaps sound waves disturb the air molecules, which precipitates the entity's appearance.

In view of these requirements, it becomes clear why this phenomenon is seldom seen in the conventional experimental laboratories. Philosophically, scientists prefer clinically clean and highly uniform conditions; "dirty" experiments that introduce too many variables are avoided. However, the natural environmental conditions should be present if anomalies are to appear.

Historic entity experiments

I now provide three specific and quite fascinating cases where these self-contained entities have been produced artificially. The equipment required for these experiments is neither exotic nor expensive.

The first of these experiments appeared in the *Electrical Experimenter* (see Fig. 18-7): In Weisiger's and Leduc's experiments, the sharp-pointed electrodes and the semiconducting surface of the film, mentioned in item three of the requirements, are notable factors.

Another example of these experiments is from the journal, *Science*, n.s. 1910, (see Fig. 18-8): A.T. Jones' report, a very simple short-circuit experiment, is at least partly covered by item #3. When duplicating experiments using short circuits, the condition of the wire—if it is cut or bent near the contact point—will alter the results. Unfortunately, there is no mention of moisture or oxides here, just of the tiny 1-mm copper ball left behind. The color of the ball should be considered indefinite, since Jones expressed some doubt.

My last and most interesting report, every bit as important as Franklin's kite experiment, is taken from two rare sources: *De L'electricite des Meteores* (Volume 1) by Professor Abbé Bertholon (1787). The second source is *A Complete Treatise on Electricity in Theory and Practice; with Original Experiments* (Volume II), by Tiberius Cavallo, (1795).

UNUSUAL DAMAGE BY A TORNADO

By E. S. DAVY
Assistant Director, Royal Alfred Observatory, Mauritius

The island of Mauritius in the South Indian Ocean is in a region in which tropical depressions or "cyclones" frequently develop and move; small tornadoes or whirlwinds are, however, rarely reported here.

On May 24, 1948, at Curepipe, Mauritius (approximately 1,850 feet above sea level), a phenomenon of the tornado type exhibited a very interesting feature which the writer has not seen mentioned in reports of other tornadoes. During most of the day the weather was mainly fair at Curepipe and convection appeared to be no more intensive than is normal for this region. In the middle of the afternoon a whirlwind touched down to earth over a hard tennis court, causing considerable damage to its hard compact surface. A trench running in a north-south direction, 60 feel long and 1 to 2$^1/_2$ feet wide, was cut in the bare surface of the court to a depth varying from 1 to 4 inches. The material lifted from the trench was all thrown to the west to a distance of 50 feet; pieces weighing about one pound were thrown as far as 30 feet. The surface material was slightly blackened as if by heating, and a crackling like that of a sugar-cane fire was heard for two or three minutes. The court was made of a ferruginous clay, which packs down to a surface more smooth than that of the hard tennis courts usually made in Great Britain.

Unfortunately, there were no very reliable witnesses of the phenomenon. The impressions gained by two servants of the tennis club who saw the incident differed considerably on important details. One claims to have seen a ball of fire about two feet in diameter which crossed from a football pitch to the tennis court through a wire-netting fence without leaving any evidence of its passage until it bounced along the court, making the trench in the surface before disappearing completely.

The high winds and ascending currents in this small tornado were unusually irregular in their action; on the tennis court an umpire's chair weighing about 50 pounds was picked up from 50 yards to the west of the trench, carried upwards to an estimated height of 60 feet, said to have been broken whilst in the air, and dropped in pieces about 20 feet to the east side of the trench, that is, on the side opposite to the pieces of court surface. A hundred yards to the south of the tennis court the roof of a small building was lifted off and dropped nearby. About two miles to the north a rather insecure building was blown over as the whirlwind moved in a northerly direction. No other evidence of damage to buildings or vegetation was reported nor could any be seen from a hill-top which was in the track of the phenomenon, between the damaged sites of which a good view was obtained.

18-6 Unusual damage by a tornado. (*Weather*, 1949). Thanks to Royal Meteorological Society for reprint permission.

I retyped the second account because of the poor quality of the paper and the hard-to-read script common to eighteenth-century writings, The Sketch (see Fig. 18-9) is found in Bertholon's description.

"Ball Lightning" Experiments
By Samuel S. Weisiger, Jr.

In the January, 1916, issue of the *electrical experimenter* you publisht a discourse on "Ball Lightning," and gave instructions for the experimental production of it. Thru the kindness of Mr. Porter, Instructor in Physics at the Allegheny High School, I have been able to make the several photos accompanying this letter. Under each photo there is given a short description of the circumstances under which each discharge was made and the phenomena connected therewith.

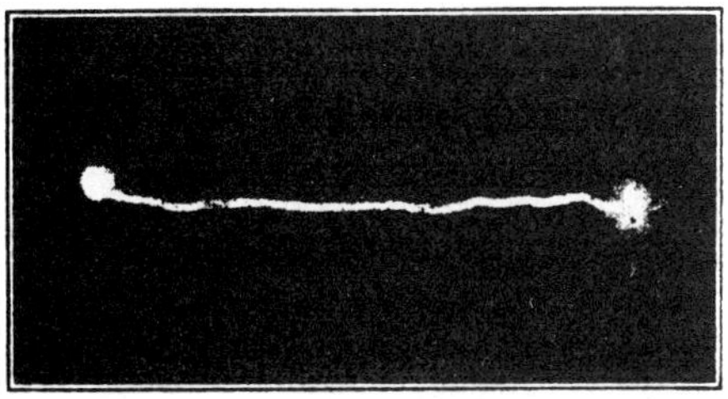

This is Another "Freak" Discharge. The Ball Travelled in a Very Crooked Path to the Positive Electrode, and Here Exploded. The Force of the Explosion Was So Great that a Part of the Spark-Ball Was Thrown to the Other Side of the Positive Electrode, from Whence It Continued to the Positive Electrode.

In making these photos a 75,000 volt Toepler-Holtz static machine was used. The distance between the sharp metal points was from 5.5 to 6 centimeters. This distance must be found by experiment, and altho it is absolutely essential to have the correct distance between points, it will nevertheless differ with the capacity of the static machine.

Much trouble will be encountered if the sharp points, used to produce the discharge are not free from grease and highly polished. The best way to polish the points is to take a little powdered chalk (blackboard chalk which has been scraped to a fine powder with a knife) and put it on some kind of cloth and turn the point of the electrode, at the same time giving considerable pressure to the cloth where the point is being turned.

The best connection for the electrodes was found to be obtained by means of two brass chains.

Two large-sized sharply pointed darning needles suitably mounted form admirable electrodes. It is practically impossible to use blunt needles.

There will be much trouble in finding the correct spacing for the electrodes and it will probably require some experimentation. In any case the spacing is dependent on the power of the static machine.

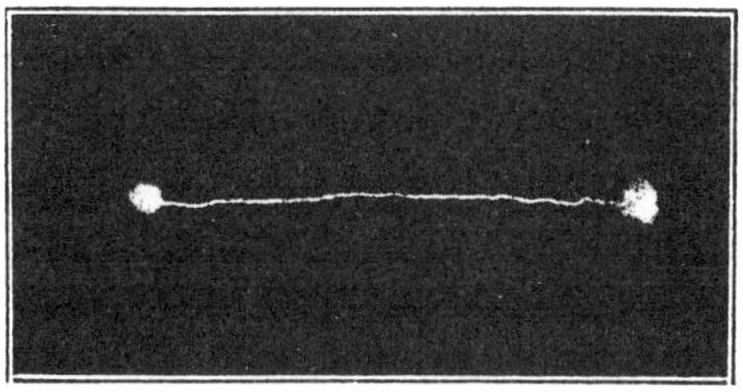

Some Trouble Was Encountered in Getting this Spark-Ball to Form. Evidence of This is Shown by the Plate Being Exposed By a Tiny Charge Or Burst of Light On One Side of the Negative Electrode. The Uneven Course of the Spark-Ball is Clearly Defined.

When the plate is put under the electrodes be sure to get the emulsion side up as the discharge occurs better when the plate is placed in this manner.

When the plate is under the electrodes and the static machine has been started, the spark ball should form very quickly. After the ball has detached itself from the electrode, turn the machine very slowly in order to expose the plate longer. The rate of travel of the spark ball is proportional to the speed of the static machine.

Should the machine be stopt before the spark ball reaches the other electrode, the plate will only show the path of the ball to that point.

Knowing that there is considerable interest in these "Ball Lightning" experiments we have repulbisht below the original directions for producing ball lightning in the laboratory as outlined by the famous French scientist—M. Stephane Leduc. His experiment makes possible the production of a slowly moving globular spark not easily obtainable in any other way, in so far as we know.

18-7 "Ball-lightning" experiments. *Electrical Experimenter,* 1919

To produce this imitation ball lightning it is necessary to employ two very fine highly polished metallic points, each of which is in connection with the positive and negative poles, respectively, of a static machine of small or medium size. These two metallic points must rest perpendicularly, as our illustration indicates, on the sensitive face of a gelatin bromid of silver photographic plate, which is placed on a metallic leaf, such as tinfoil. The two metal points are spaced about five to ten centimeters apart. When the static machine is operated an effluvium is produced around the positive point, while at the negative point there is formed a luminous *fireball* or *globule*.

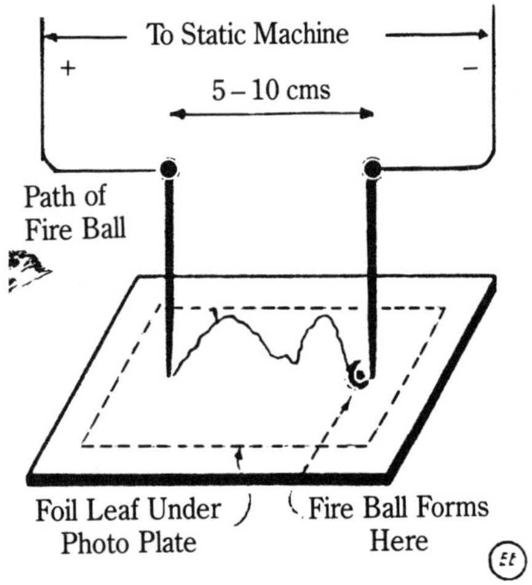

Scheme for Producing Ball Lightning in the Laboratory with Static Machine. Photograph Plate and Two Needles.

Now, when this globule has reached a sufficient size, it will be seen to detach itself from the metallic point, which then ceases to be luminous, and the globule will begin to move forward slowly over the surface of the plate, taking various curved paths and eventually it will set off in a direction toward the positive metal point. When it reaches this electrode the effluvium is extinguished and all luminous phenomena ceases. Further, the static machine acts as if its two poles were short-circuited, or, in other words, united by a conductor.

The velocity acquired by the luminous globule as it travels is quite slight, it taking from one to four minutes for it to traverse a path of six centimeters in some cases, and before reaching the positive electrode the globe bursts into two or more luminous balls which individually continue their journey to the positive electrode. On developing the photographic plate (which, of course, should be placed under a ruby light while the foregoing experiment is conducted) there will be found a trace on it of the exact route followed by the spark globule—the point of explosion, the routes resulting from the division, and the effluvium around the positive electrode point.

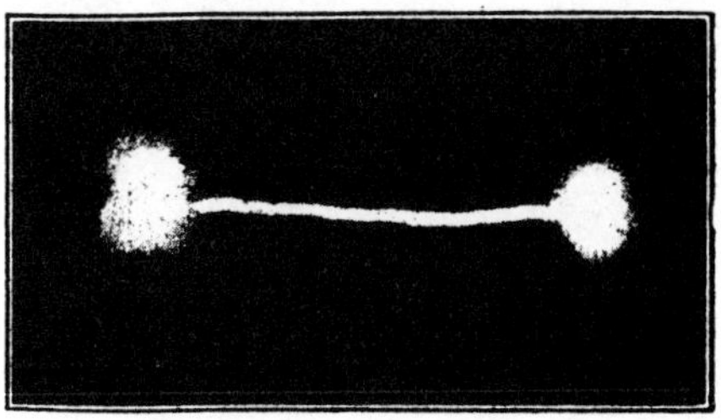

This is Probably the Best Photo of the Set, the Spark-Ball Being the Largest Obtained. You Will Notice the Manner in Which the Ball Broke Into Two Parts and Each Part Proceeded to the Pole. The Effluvium Around the Positive Pole Shows Signs of a Violent Explosion As Will Be Noted by Closely Examining the Tree Formation Made By the Bursting Spark-Ball.

Also, if one should stop the experiment before the globule's arrival at the positive electrode, the photograph will only give the route to that point. The fireball takes for its course the conductor, which apparently short-circuits the static machine. If sulfur or some other powder is thrown on the photographic plate while the experiment is being conducted, and also while the ball is moving, its path will be marked by a line of aigrettes, looking very much like a luminous rosary.

[*The Editors will be glad to hear from any of our readers who have made experiments in this direction. Photographs are particularly welcome.*—Ed.]

18-7 Continued

A LABORATORY ILLUSTRATION OF BALL LIGHTNING

In Dr. Elihu Thomson's address at the opening of the Palmer Physical Laboratory at Princeton University he made, with regard to ball lightning, the statement, "The difficulty here is that it is too accidental and rare for consistent study, and we have not as yet any laboratory phenomena which resemble it closely." This suggested to me that a phenomenon which I witnessed some six or seven years ago might be worth recording.

With a copper wire a student accidentally short-circuited the terminals of an ordinary 110-volt circuit. I happened at the time to be a few meters from him and to be looking toward the terminals. At the instant of the short circuit I saw an incandescent ball which appeared to roll rather slowly from the terminals across the laboratory table and then disappeared. As I remember it, I should say that the ball may have appeared to be about three centimeters in diameter. I think no one else in the room saw anything more than a flash of light—much as if a fuse had blown. On the table where the ball had rolled we found a line of scorched spots, as if the ball had bounced along the table and had scorched the wood wherever it touched. As I remember them, these scorched spots were rather close together, perhaps not more than one or two centimeters apart. In the top of the table was a crack perhaps a millimeter or two wide, and at this crack the scorched line ended. In a drawer immediately under this crack we found a tiny copper ball, perhaps a millimeter in diameter. Apparently the ball that rolled along the table was incandescent copper vapor, although my memory of it is rather of a yellow-white than of a greenish light.

The above suggested the possibility of a laboratory study of a phenomenon which may very possibly be similar to that of ball lightning, but I have never attempted to repeat the experiment.

A. T. Jones
Purdue University

18-8 A laboratory illustration of ball lightning. *Science*, n.s. 1910

EXTRACT OF A LETTER FROM MR. ARDEN, LECTURER, IN NATURAL PHILOSOPHY, DATED SEPTEMBER 25, 1772

"About fourteen or fifteen years ago, in the presence of Wm. Constable, Esq; at his seat at Burton Constable, in Holderness, I made the following experiments:

"I placed a large coated jar, that would hold three or four gallons, directly under the prime Conductor of a very good electrical machine. The prime Conductor was at least eight or ten inches above the top of the jar, and the communication was made by a brass wire, bent at one end over the prime Conductor, and the other end passed through a small glass tube (contrived by Mr. Constable to prevent the electric matter from easily flying off) was suspended in the middle of the jar, and had a small piece of brass chain fastened to it, that rested on the bottom of the jar.

"I then began to turn the wheel, and, after turning about 100 or 150 times, as low in the jar as I could see for the coating, I perceived a ball of fire, much resembling a red-hot iron bullet, and full three quarters of an inch in diameter, turning round upon its axis, and ascending up the glass tube that contained the brass wire, which was the Conductor to the inside of the jar.

"I immediately asked Mr. Constable, if he saw the ball of fire? he said, Certainly. I said, I will turn on. He answered, By all means. I kept turning the wheel, and the ball of fire continued turning upon its axis, and ascending up the glass tube till it got quite upon the top of the prime Conductor. There it turned upon its axis some little time, and then gradually descended, turning upon its axis as it had done in its ascent, and so continued till it was so much below the top of the coating that we could no longer see it. But soon

18-9 The Arden and Constable experiment. *De L' Electricite' Des Meteores*, Vol. 1, 1787, Abbe Bertholon

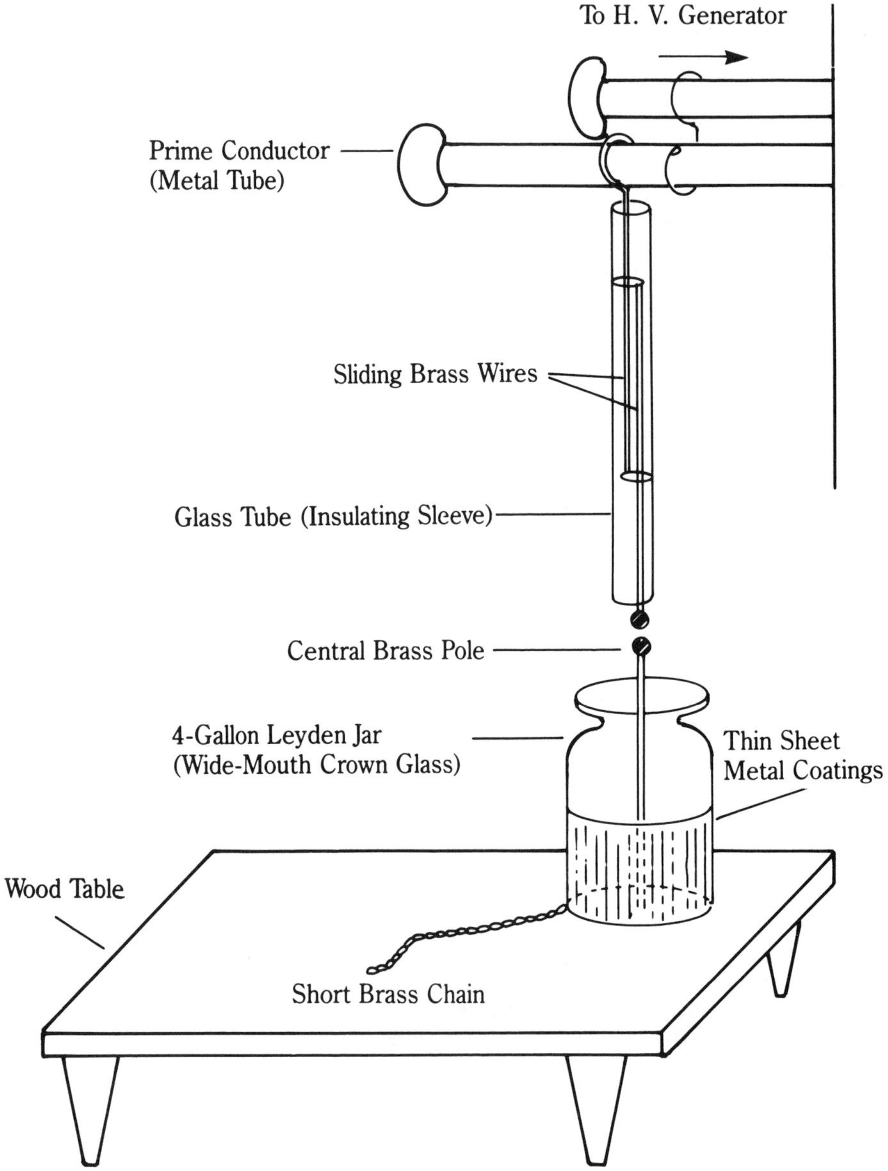

after this, a very great flash was seen; a large explosion was heard, and strong smell of sulfur was perceived all over the room; a round aperture was cut through the side of the jar, as fine as if it had been cut with a diamond, rather more than three quarters of an inch in diameter, and between two and three inches below the top of the coating, and the coating was torn off all round the aperture, about three or four inches in diameter. The jar was a pretty strong one, of crown glass.

18-9 Continued

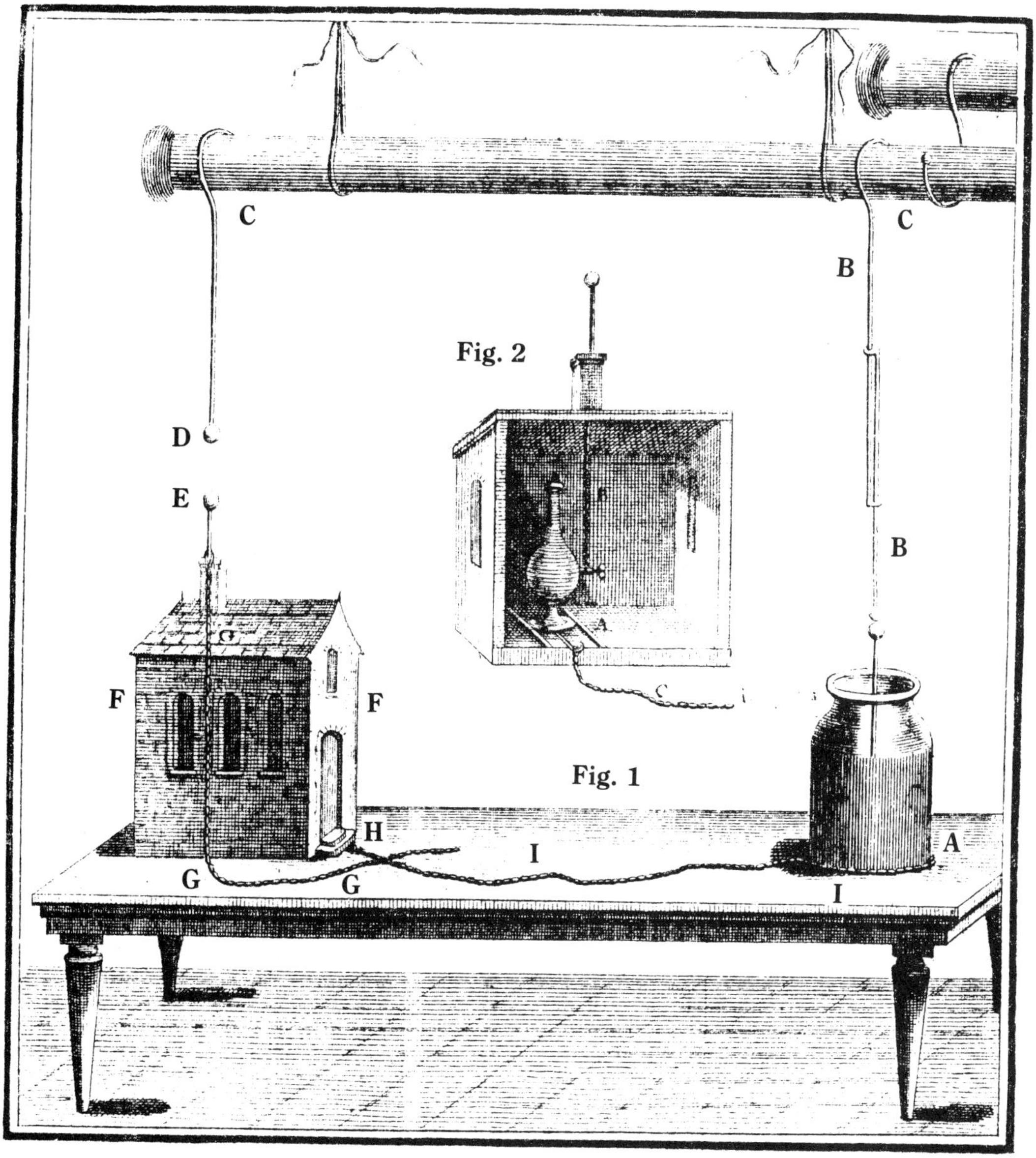

"I then took another jar, so like the first, that when both were whole I could not easily perceive any difference between them. I then attempted to charge this jar, in the same manner as the other, and we both observed it very accurately. No ball of fire was seen, but presently the jar discharged itself with a great flash and explosion, and at about the same part as of the first jar; but instead of the aperture which was made in the first jar, there was a circle about three quarters of an inch diameter, as white as chalk, and the coating torn off round about it as before. Upon touching the white part, it dropped out, and appeared to be glass in a fine powder.

18-9 Continued

"We broke several other different-sized jars that day, (which made Mr. Constable say we were in great luck) but without any thing else remarkable.

"The first experiment was made soon in the afternoon of a clear day, and the machine stood directly between us and a window, which was not above a yard from it. I don't hear that the ball of fire has been produced by art by any one else, to this day, although it is often produced by nature.

"I had the pleasure of seeing Mr. Constable this day, and of reading the account of these experiments to him, and, to the best of his memory, he thought the whole was strictly true.

"Mr. Constable thinks it would not be difficult to repeat the experiment, and to produce the ball of fire at any time, provided the jar is large, and not coated too near the top, and that the wire communicating from the prime Conductor to the inside of the jar is made to pass through a small glass tube (which is certainly of great advantage of making experiments of this kind) and that the machine acts very strong. If not, it will be in vain to attempt it."

18-9 Continued

I can only conclude that Mr. Constable was a man of few words!

Several key points are worth noting in Arden and Constable's experiments: the Leyden jar's central wire has several points of loose contact, including chain links and a sliding joint. The jar is large and is charged slowly, almost to the flash-over point. The open jar, the overhead window, and the fact that the visible fireball was formed early in the tests, but not later, indicates that a film of moisture in the glass jar and the central glass tube is the most important requirements. In the historical evolution of the Leyden jar, the early form included no lid. Natural philosophers soon discovered that by blowing into the jar, the moisture would greatly increase the storable charge.

Many theories have been advanced to account for the large amounts of energy and long lifetimes of these electric entities (usually 1 to 5 seconds). Fireballs have been known to cause major damage to tile roofs and chimneys, bend heavy iron gates and door hinges, bring a barrel of water to its boiling point in a short time, and bore small holes through granite blocks. Some of these points are described in Brand's report of 1923. An especially good book on the subject is *Ball Lightning and Bead Lightning* by James D. Barry (1980). None of the theories mentioned in these two sources require a major shift in fundamental physics concepts.

Theoretical implications

I personally feel these explanations do not account for the enormous energy content in a small space. The three artificial electric-entity productions featured here involve no large power requirements to initiate a formation.

Following are two unorthodox theories, seldom seen:

1. In line with Gustave Le Bon's ideas on the universal dissociation of matter, a fireball would be seen as a slow energy-conversion process, in which elemental matter (water molecules, for example) yields its intrinsic potential energy. When the quantity of moisture is slowly reduced to a low level, the fireball becomes starved out of existence. Le Bon pointed out the large number of electrons bound in a single gram of water. Should these electrons be freed to appear as charge, an enormous 96,000 coulombs of electricity would be produced. One coulomb applied to each of two spheres 1 meter apart represents an electrostatic force of 2 billion pounds! Normally, only tiny fractions of a coulomb are found in nature.
2. The second view involves a return to our discussion of gravitation. In this theory, an energy conversion occurs that disturbs the "concealed or hidden" motions in space (hypothesized by Heinrich Hertz in 1899 to account for the storage of potential energy).

Also, odd gravity anomalies are associated with electric entities, specifically, heavy, fragile objects that fall without being broken. This action implies that the basic properties or constants of space itself have been altered. Since the true nature of electrification is concealed at the molecular level, perhaps electrification produces disturbances in the incessant, hidden motions of space. These disturbances are manifested as heat, light, and mass motion—one form of which is the appearance of ball lightning.

These two views, of course, are quite unsettling because they require a major paradigm shift in fundamental physics concepts and an enlargement of our scientific foundational principles. Some physicists have admitted that the formation of ball lightning in metal enclosures, such as airplanes, raises the question of how such large energy densities in a small space are maintained (see *Nature*, volume 224, 1969, p. 895, for example). The metal enclosure ensures that fireballs are not supplied externally with electromagnetic energy, as some theoretical physicists have imagined. We have, in ball lightning, a means by which energy is extracted from the nearby quiescent environment and manifests itself as heat, light, and mass-motion. If this conclusion is justified, then it would be a natural case of *negative entropy*; that is, energy flowing "uphill," not downhill, as is required by the Second Law of Thermodynamics.

One article in favor of this limitation is "The Second Law of Thermodynamics and the 'Death' of Energy, with Notes on the Thermodynamics of the Atmosphere," by Charles P. Steinmetz from the *General Electric Review* (July 1912).

> Expressing the second law of thermodynamics in the words: "Without expenditure of some other form of energy heat flows only from higher to lower temperature," the author shows that the logical sequence from this is the conclusion that eventually all energy transformation will stop, i.e., all motion will cease and the universe will be dead. The conclusion is not a reasonable one and the author sets out to disprove the general applicability of the law. Adopting as his line of reasoning the thermodynamics of gases, he shows how, attending the escape of molecules from the attraction of earth into cosmic space, there is a heat energy flow from a temperature of 10 deg. C. to one of 60,000 deg. C. Even within the earth's atmosphere, and without considering what happens in cosmic space, he shows that there is a transference of heat energy from lower to higher temperatures, or rather against the thermodynamic temperature equilibrium; and leads us to the conclusion that this law of thermodynamics is not of universal application, but applies only within the limited range of thermodynamic engines, from which it has been derived.

In addition to Dr. Steinmetz, physicists James Clerk Maxwell, Thomas Preston, and Lucien Poincare held a similar view of the Second Law of Thermodynamics.

A thought-provoking implication presented itself when I reflected on Arden and Constable's simple experiment, performed during the latter part of the eighteenth century. With their homemade frictional generator and Leyden jar, they succeeded in producing a phenomenon that today's government research centers have failed to duplicate using the best high-powered generators and large financial investments.

Extrapolating from this paradox, it now appears that it is possible to produce quite anomalous results by employing a larger number of principles through which nature operates. The experimental lab should duplicate natural environmental conditions; sterility and uniformity are often barriers to the discovery of new natural laws.

The application of this philosophical approach could result in a great simplification of our technology, making it more reliable with less waste and pollution as by-products. These possibilities require a greater resiliency and a willingness to think in different modes, some of which have not been emphasized in our present educational system. Fortunately, the spirit of inquiry and creativity is innate in each new generation of children.

19 Some philosophical conclusions and insights

AS INTERESTING AS FOREGOING EXPERIMENTAL AVENUES MIGHT SEEM, the most important part of the book is discussed in this chapter. That is, how we see, think, and reason on scientific subjects is more important than project plans. The character of our science and technology, especially as it impinges on the environment in the looming specter of toxic wastes and the ravaging of earth's resources, should impel us to look for lasting solutions.

Seldom do we go far enough to reexamine how we think through solutions without creating new problems in their wake. I previously mentioned the importance of intuition and visualization as elements of the innovative process. The three main elements that are important in creative thinking are: visualization, intuition, and qualitative anatomization.

Visualization

By *visualization* I am not referring to a self-hypnotic state, but rather that ability to form models in thought, as for example, the Le Sage concept of gravitation.

In Newton's time, there was an increasing fascination with abstract mathematics as a method of "seeing" scientific principles. By the turn of the twentieth century, visualization as a tool had fallen into disuse.

In one sense, abstract mathematical reasoning is identical to taking a subway in a large city; we have no way of knowing whether a far shorter route exists. Now, please do not conclude from what I have said that mathematics and its offspring, statistical and computer analysis, have no place in scientific innovation. Rather, they must be balanced with an equal emphasis on the other ways of "seeing" the way in problem solving.

Generally, mathematics should be used to find exact solutions only after we have diligently developed working mental models to explain phenomena. Finding these mental models requires participating in concrete laboratory experiments. The use of math without physical experiments leads to the construction of "air castles," which have no firm foundation in reality. Ultimately, you must visualize the principle under consideration before you can use the full powers of your intellect to find an innovative solution.

Intuition

Intuition is the ability to reach a solution without recourse to reasoning or inference. It is an inner knowing or deep feeling. Even though intuition is innate in children, it usually is suppressed by the educational system.

Although logical, linear thinking proceeds from point A to point B to point C; intuitive thinking leaps from A directly to C. Sometimes intuition is thought to be the exclusive domain of women's thinking; this idea is completely unfounded. For example, inventor Nikola Tesla, who pioneered much of our existing electrical distribution system, often had flashes of insight that supplemented his mathematical reasoning.

The long history of inventiveness in the United States has not taken gradual steps, but usually has been proceeded by intuitive leaps. For example, powered flight and the development of the telephone come to mind.

Now consider some aspects of contemporary technology and compare them to what our intuitive sense indicates they should be. Technical devices such as motor vehicles could be thought of as noisy, considerably inefficient, polluting and complex. What should they be? In this case, we can arrive at the answer almost by inversion: silent, efficient, nonpolluting, and very simple (reliable). So now we can assess what are truly innovations in transportation. Normally, intuition answers our questions of what should be.

Qualitative anatomization

Qualitative anatomization is a method of seeing that involves a resolution of the object under consideration into one-word qualities that characterize the object. The following are three everyday examples of qualitative anatomization. The words in parentheses simply explain our quality choice.

Pencil:
Communication
Rigidity

Utility Pole:
Communication (supports phone lines)
Rigidity
Uniformity
Sterility (nonliving)
Safety (supports and protects power lines)
Supply (links to electrical energy)

You would hardly expect to find a landscape painter getting ecstatic over the prospect of painting a panorama of utility poles. Why? Because uniformity and sterility are not attractive qualities in landscape.

Tree:
Communication (for flocks of birds)
Rigidity (main trunk and limbs)
Permanence (settled, rooted)
Protection (for birds and squirrels)
Individuality
Resilience (twigs and leaves)
Supply (leaf nutrients and lumber)
Comfort (natural air-conditioner)
Beauty
Alive

Note that trees are a "higher" expression because they include more qualities; the lesser (pencils and poles) can only come from the greater (trees).

Qualitative anatomization is not a trivial pursuit. It exercises those mental faculties that we need to order to "see," just as poets and painters see their surroundings. With continued practice, this tool becomes very helpful for assessing the value of both objects and abstract concepts, such as beauty or work. Qualitative anatomization also assists our experimental work. Now, combine these three aspects of creative thinking into a mental juggling act. The list of qualities you arrive at will be strongly influenced by your reference point. For example, your point of view might be influenced by possible social or environmental impacts.

The reference point concept can be illustrated by contrasting the nature of progress, as it is now, with what your intuitive sense believes it should be. The viewpoint is that of social impact; how progress relates to people.

Progress

Contemporary	*Intuitive Sense*
Complexity	Simplicity
Specialization	Diversity
Self-Pollution	Nondestructive
Material acquisition	Spiritual wealth
Rigidity	Resilience
Without identity	Individualism
Narrowing of skills	Multiple talents, skills
Sterile uniformity	Spontaneous variety
Vulnerability	Stability, strength
Centralized control	Local autonomy
Transience, homelessness	Settled, rooted
Convenience, ease	Nurturing, individual responsibility
Insecurity, stress	Peace of mind
Frenzy, rush	Poise, dignity

When we hold these contrasting views of progress in mind, we can then assess which new developments in technology truly lead us into a better world. Science and technology must be responsible to social needs, and to how their products affect the character of people.

If we would give serious attention to these modes of thinking, it would enable the rising generation of students to work with qualities as precisely as they deal with mathematics. Qualification would then reach an equal standing with quantification.

Some useful surroundings

A natural setting is helpful for reflective thought. Surrounded by nature's examples, we are more likely to recognize different principles that operate in the environment. Silence and a relaxed, tranquil attitude are essential.

At first, strongly focus on the technical problem to be solved. If answers and insights do not come quickly, put the task out of your thoughts entirely. The worst mistake is to concentrate on a problem until it is solved. This is typical of the ''think tank'' or ''brain trust'' approach; it usually leads to a myopic, hypnotic state, in which you cannot recognize new principles that will greatly simplify the solution.

One extreme case of think-tank hypnosis occurred to me. In 1962, I was working on a mechanics problem. After two days of thinking and drawing, I still had no answer. Eventually, I forgot about the subject until I came across it again in 1986. I briefly reviewed some aspects of the mechanics involved and made a list. Then one night, just before falling asleep, a mental picture flashed through my mind, showing the mechanism in action. I jumped up and drew a sketch, and it did satisfy all the requirements as a solution! Apparently, an incubation period is sometimes necessary. Always have a pencil and notepad on your nightstand so you don't lose these flashes of insight.

To take part in a new renaissance, men and women need to avoid becoming specialists. We should adopt the multidisciplined approach with some knowledge in several different areas. The development of manual skills is also essential in experimental research.

Orthodox science too often takes on an air of omniscience; that is, that all the basic knowledge we have is all there is. This position is further complicated by its companion attitude—infallibility. Scientists often believe that there is only one way and we have never taken wrong or dead-end paths.

Especially since the development of the hydrogen bomb in the 1940s, science and technology have taken on what might be called a "swaggering arrogance"; this attitude has no place in the science of the twenty-first century. Nature is not to be dominated. Its many principles should be copied in the applications of our sciences and technology.

Any scientific body of knowledge that has excluded natural anomalies because they do not fit with accepted notions, must not only be suspect, but be incomplete. Most of the natural anomalies found in scientific journals over the past two centuries still have not been classified and catalogued. The task for those researching the full history of scientific discovery remains awesome.

I have tried in this book to select anomalies and innovations found before the turn of the twentieth century, in the hope of bringing respect and admiration for the spirit of inquiry in those earlier times. We have much to relearn and reincorporate into our own time. We must blend it with the best of what we have to offer.

In view of this, it appears that the real "high frontier" is still very much on earth, rather than in space. The challenge is to not only reexamine the way we think, but to review the hidden assumptions in the foundation on which today's science rests.

Happy experimenting!

Appendix

Cataloguers of nature's anomalies

Fortean Times, 96 Mansfield Rd., London, NW3 2HX, England.
Prints publications on Charles Fort's research.

Pursuit (Journal), c/o Edward Brother's, Inc., 2500 S. State St., Ann Arbor, MI 48104.
A journal dealing with unexplained natural phenomena.

The Source Book Project, P.O.Box 107, Glen Arm, MD 21057.
Provides high-quality books on anomalies and publishes a newsletter *Science Frontiers*.

List of sources of materials for experimenters

American Machine & Tool Co., Fourth Ave. and Spring St., Royersford, PA 19468.
Supplies woodworking tools.

Bogert and Hooper, Inc., 175 West Carver St., Huntington, NY 11743.
Supplies wooden balls, rings, and eggs.

Burden's Surplus Center, P.O. Box 82209, Lincoln, NE 68501-2209.
Supplies 12-volt electric motors, transformers, rectifiers.

Cherry Tree Toys, Inc., Belmont, OH 43718.
Supplies wooden toy wheels and wood balls for Wimshurst generator construction.

Component Parts, 230 Hampton Parkway, Kenmore, NY 14217.
Supplies Van de Graaff generators and related apparatus, kits, and information.

Edmund Scientific Co., 101 E. Gloucester Pike, Barrington, NJ 08007-1380.
Sells lenses and general science fair materials.

Fairmount Abrasive Systems, P.O. Box 236, Wedron, IL 60557.
Supplies #46, #60, #100 aluminum oxide grains for electrostatic motor rotor experiments.

Hagenow Laboratories, Inc., 1302 Washington St., Manitowoc, WI 54220.
Supplies sulphur, paraffin oil, and chemicals.

Arthur Harris Company, 210 N. Aberdeen St., Chicago, IL 60607.
Supplies #304 stainless-steel float balls for high-voltage terminals. Has minimum order requirement.

House of Brilliance, 63 Dwight Street, New Britain, CT 06051.
Supplies minerals for cold-light research.

Hygenic Corporation, 1245 Home Ave., Akron, OH 44310.
Supplies pure latex "Dental Dam" in a 1-square yard, 6-inch roll in "heavy" and "extra heavy" thicknesses for electrophorus experiments. Write for your nearest distributor.

J & L Industrial Supply, 19339 Glenmore, Detroit, MI 48240.
Supplies tools for the wood machinist, drill bits, reamers, deburring, etc.

Jerryco, Inc., 601 Linden Place, Evanston, IL 60202.
Excellent supplier of surplus equipment, including 12-volt motors, ball bearings, optics, flash rocks, low-voltage rectifiers, transformers, and science-fair items.

N.T.E. Electronics, Inc., 44 Ferrand St., Bloomfield, NJ 07003.
Makes high-voltage diodes used in industrial and microwave ovens. Write for your local distributor.

Old World Art, 1953 South Lake Place, Ontario, CA 91761.
Sells composition gold leaf. Write for local distributor.

Precision Brand Products, 2252 Curtiss St., Downers Grove, IL 60515.
Supplies steel and brass shim stock. Write for local distributor.

Strobelite Company, Inc., 430 West 14th St., Room 507, New York, NY 10014.
Supplies luminous and fluorescent paints.

Consult the *Thomas Register* at your local library for the following: Steel and brass shim stock (for neutralizers and Leyden jars), wax, carnauba and gum, and ester (for electrophorus cakes).

United States Plastics Corp., 1390 Neubrecht Rd., Lima, OH 45801.
Supplies general acrylic plastics and clear Butyrate shipping tubes for making Wimshurst Leyden jars.

Woodcraft, P.O. Box 4000, Woburn, MA 01888.
Supplies woodworking tools, shellac flakes, and varnishes.

Useful books

De Cristoforo, R. J. *De Cristoforo's Complete Book of Power Tools*. Popular Science Publishing Co., NY, 1972.
Covers woodworking techniques using simple, inexpensive jigs and fixtures.

Stong, C. L. *The Scientific American Book of Projects for the Amateur Scientist*. Simon & Schuster, NY, 1960.

Strong, John. *Procedures In Experimental Physics*. Prentice Hall, NY, 1938.

Research Bibliography (By Subject)

Electronic equipment protection

Greason, William D. *Electrostatic Damage in Electronics: Devices and Systems*. Research Studies Press, Hertfordshire, England, 1987.

Lacy, Edward A. *Protecting Electronic Equipment From Electrostatic Discharge*. TAB Books, Blue Ridge Summit, PA, 1984.

Frictional electric generators

Dibner, Bern. *Early Electrical Machines*. The Burndy Library, Norwalk, CT, 1957.

English Mechanic. ''Winter's Electrical Machine.'' Vol. 2, Oct. 20, 1865.

Hackmann, W.D. *Electricity From Glass: The History of the Frictional Electric Machine, 1600-1850*. Sijthoff and Noordhoff, Netherlands, 1978.

Harris, William Snow. *A Treatise on Frictional Electricity*. Virtue and Co., London, 1867.

Weinhold, Adolf. *Introduction to Experimental Physics*. Longmans, Green and Co., London, 1875.

The influence machine (and modifications)

British Patent No. 22, 731, Tudsbury.

Direct Current. "Electrostatic Generators." Noel J. Felici, Vol. 1, June 1953.

The *Electrician*. "Influence Machines." Vol. 35, London, 1895.

Gray, John. *Electrical Influence Machines: Their Historical Development and Modern Forms*. Whittaker & Co., New York, 1903.

Johnson, Valentine E. *Modern High Speed Influence Machines*. E. and F.N. Spon, London, 1921.

Jolivet, Pierre. *Sur une novelle machine Electrostatique a Influence*. R.G.E., Paris, France, 1953.

Journal of the Rontgen Society. "The Wommelsdorf Condenser Machine." Jan. 1914.

Journal of the Telegraph Engineers. "The Influence Machine From 1788 to 1888." Vol. 17, Silvanus P. Thompson, London, 1888.

Society Frances Des Electriciens. "The Chaumat Electrostatic Machine." Bulletin #1, p. 673+, July 1931. (In French.)

U.S. Patent No. 882,508 and 1,071,196, Wommelsdorf.

U.S. Patent No. 634,467 and 720,711, Lemstrom.

U.S. Patent No. 937,691, Baker.

U.S. Patent No. 821, 902, Todd.

U.S. Patent No. 1,109,205, Dempster.

General electrostatics

Craggs, J.D. and J.M. Meek. *High Voltage Laboratory Technique*. Buttersworths Scientific Publ., London, 1954.

Heilbron, J.L., *Electricity in the 17th and 18th Centuries, A Study of Early Modern Physics*. University of California Press, Berkeley, CA, 1979.

Moore, A.D. *Electrostatics*. Anchor Doubleday & Co., NY, 1968.

Moore, A.D. (ed.) *Electrostatics and Its Applications*. John Wiley & Sons, New York, 1973.

United States Patent Class Listing. "Electrical Generation—Induction Type," Class #310, Subclass #309.

This is a list of approximately 300 U.S. Patents on Electrostatic Induction Devices, Including All Influence Machines. Available from the U.S. Patent Office.

The electroscope

Dolbear, A.E. *The Art of Projecting*. Lee & Shepard Publishers, Boston, 1877.

Le Bon, Gustave. *The Evolution of Forces*. Kegan Paul, Trench, Trubner & Co., London, 1908.

Le Bon, Gustave. *The Evolution of Matter*. Walter Scott Publishing Co., London, 1910.

Journal of Orgonomy. "The Electroscope (Part 1)." Vol. 3, No. 2, C. Fredrick Rosenblum, 1969.

Journal of Orgonomy. "The Electroscope (Part II)." Vol. 4, No. 1, C.F. Rosenblum, Orgonomic Publications, Inc., Princeton, NJ, 1970.

Royal Society of London, Proceedings. "C.T.R. Wilson Electroscope." Vol. 68, p. 152, 1901.

Rutherford, Ernest. *Radio-activity*. Cambridge University Press, England, 1904.

Tributsch, Helmut. *When Snakes Awake: Animals and Earthquake Prediction*. MIT Press, Cambridge, MA 1982. Discusses electrometer anomalies and earthquake lights.

Weinhold, Adolf. *Introduction to Experimental Physics*. Longmans, Green and Co., London, 1875.

The Leyden jar

Heilbron, J.L. *Electricity in the 17th and 18th Centuries, A Study of Early Modern Physics*. University of California Press, Berkeley, CA, 1979.

Popular Electricity Magazine. "Liquid Condensers." Chicago, IL, 1912-13.

The electrophorus

Harris, William Snow. *A Treatise on Frictional Electricity*. Virtue and Co., London, 1867.

Philosophical Magazine. "On a Modification of the Electrophorus." Series 3, Vol. II, John Phillips, London, 1833.

Popular Mechanics. "Mushroom Generator." New York, 1935.

Weber, Joseph. *Abhandlung Von Dem Luftelektrophor*. Ulm, Switzerland, 1779.

Research avenues

Electrostatic motors

American Journal of Physics. "Operation of Electric Motors from the Atmospheric Electric Field." Vol. 39, Oleg Jefimenko, July 1971.

Jefimenko, Oleg D. *Electrostatic Motors: Their History, Types, and Principles of Operation*. Electret Scientific Co., Star City, WV, 1973.

Popular Electricity Magazine. "Static Electric Top." Chicago, IL, 1912.

Electrohorticulture

Lemstrom, Selim. *Electricity in Agriculture and Horticulture*. The Electrician Printing & Publishing Co., London, 1904.

U.S. Patent No. 1,268,949, Fessenden.

U.S. Patent No. 1,952, 588, Golden.

Electrotherapeutics and high-voltage humans

Annals of Electricity, Magnetism & Electricity. "Extraordinary Case of Electrical Excitement." Vol. II, London, pp. 351-354, 1838.

Becker, Robert and Gary Selden. *The Body Electric*. William Morrow, NY, 1985. Deals with general subject of bioelectricity.

Coulter, Harris L. *Divided Legacy: A History of the Schism in Medical Thought, Vol. III. Science and Ethics in American Medicine 1800-1914.* McGrath Publishing Co., Washington, DC, 1973.

Electrical Experimenter. "Electrified Convicts." Vol. 8, New York, p. 185, 1920.

New York Times. "Electrified Convicts." p. 1, Monday, April 5, 1920.

Strong, Fredrick Finch, M.D. *Essentials of Modern Electro-Therapeutics.* Rebman Co., New York, 1908.

Cold light

Birkeland, Kristian. *On the Cause of Magnetic Storms.* Longmans, Green & Co., New York, 1908.

Corliss, William R. *Lightning, Auroras, Nocturnal Lights and Related Luminous Phenomena.* The Source book Project, Glen Arm, MD, 1982.

Crookes, William. *Radiant Matter.* James Queen & Co., Philadelphia, PA, 1881.

Journal of the Optical Society of America. "Triboluminescence." Vol. 29, Wisconsin, p. 407+.

Le Bon, Gustave. *The Evolution of Forces.* Kegan Paul, Trench, Trubner & Co., London, 1908.

Meteorological Magazine. "Lemstrom's Aurora." pp. 33-36, 51-55, London, April 1883.

Physical Review. "The Electron Theory of Phosphorescence." Series 2, Vol. 1, Chester Butman, 1913.

Science, n.s. "Lemstrom's Aurora." Vol. 4, pp. 465-466, 1884.

Scientific American. "Puluj; Phosphorescent Lamp." Vol. 77, New York, 1897.

Electroaerodynamics

Dudley, Horace C. *Analog Science Fact & Fiction.* "The Electric Field Rocket," November 1960.

Product Engineering. "Sonic Boom Experiments." Vol. 39, New York, pp. 35-6, March 11, 1968.

U.S. Patent No. 3,095,167, Dudley.

Counter-gravitation

American Philosophical Society, Proceedings. Philadelphia, PA, years 1914-1929. See several articles on Charles F. Brush's experiments.

Electrical Experimenter. "Can Electricity Destroy Gravitation?" New York, March 1918.

Electrical Experimenter. "Piggott's Electro-gravitation Experiment." Vol. 8, 1920.

Hooper, William J., *New Horizons in Electric, Magnetic and Gravitational Field Theory*, Principia College, Elsah, IL, 1974.

The Scientific Papers of James Clerk Maxwell. Vol. II, W.D. Niven (ed.), Constable & Co., London, 1965. See Maxwell's "Le Sage Theory of Gravitation."

Transactions of the Academy of Science. "Nipher's Gravitation Experiments." Vol. 23, pp. 163-192+ St. Louis 1916.

U.S. patent no. 1,006,786, Piggott; 3,518,462, Brown; 3,610,971, Hooper.

See, Thomas Jefferson, *WAVE THEORY! Discovery of the Cause of Gravitation!* self-published, London, 1938.

Science and Invention. "Controlling Gravitation." New York, August, 1929.

Unusual electrical discharges

Barry, James D. *Ball Lightning and Bead Lightning: Extreme Forms of Atmospheric Electricity*. Plenum Press, New York, 1980.

Brand, Walther. *Der Kugelblitz* (Ball Lightning). Hamburg, Germany, 1923. (NASA Translation: NASA T.T. F-13 228, year 1971, Accession Number N71-18133.)

Cavallo, Tiberius. *A Complete Treatise on Electricity In Theory and Practice with Original Experiments*. Vol. II, 1795.

Corliss, William R., *Lightning, Auroras...*

Corliss, William R. *Tornadoes, Dark Days, Anomalous Precipitation, and Related Weather Phenomena.* The Sourcebook Project, Glen Arm, MD, 1983.

Bertholon, Abbe. *De L'Electricite Des Meteors.* Vol. 1, 1787.

Electrical Experimenter. "Weisiger's Ball Lightning Experiments." Feb. 1919.

Ehrenberg, W. *Scientific American.* "Maxwell's Demon." Vol. 217, pp. 103-110, Nov. 1967.

Fort, Charles. *The Complete Books of Charles Fort.* Dover Publications, New York, 1975.

Journal of Applied Physics. "The Atomphysical Interpretation of Lichtenburg Figures and their Application to the Study of Gas Discharge Phenomena." Vol. 10, Dec. 1939.

Maxwell, James Clerk. *Theory of Heat.* D. Appleton & Co., NY, 1872.

Meaden, George Terence. *The Circles Effect and Its Mysteries.* Artetech Publishing Co., Wiltshire, England, 1989.

Persinger, Michael and Gyslaine Lafreniere. *Space-Time Transients and Unusual Events.* Nelson-Hall, Chicago, IL 1977.

Poincare, Lucien. *The New Physics and its Evolution.* Kegan Paul & Co., London, pp. 86-87, 1907.

Preston, Thomas. *The Theory of Heat.* MacMillian & Co., NY, 1904.

Ryan and Vonnegut. *Science* n.s. "Miniature Whirlwinds Produced in the Laboratory by High-Voltage Electrical Discharge." Vol. 168, pp. 1349-1351, June 12, 1970.

Science n.s. "Laboratory Ball Lightning." Vol. 31, p. 144, 1910.

Scientific American Supplement. "Remarkable Effects of Lightning." Vol. 57, p. 23679+, 1904.

Steinmetz, Charles P. *General Electric Review.* "The Second Law of Thermodynamics and the 'Death' of Energy, with Notes on the Thermodynamics of the Atmosphere." July 1912.

Wainwright, Jacob T. *The Engineer.* (London) "The True Second Law of Thermodynamics." Vol. 113 pp. 658-9, June 21, 1912.

Weather: A Monthly Magazine for all Interested in Meteorology. "Unusual Damage By A Tornado." Royal Meteorological Society, London, 1949.

Philosophical implications

Dreyfus, Hubert and Stuart Dreyfus. *Mind Over Machine: The Power of Human Intuition and Expertise in the Era of the Computer.* The Free Press, NY, 1986.

Elkin, Benjamin. *The Loudest Noise in the World.* Viking Press, NY, 1954.

List of associations, libraries, museums

The Bakken: A Library and Museum of Electricity in Life, 3537 Zenith Avenue South, Minneapolis, MN 55416. The Bakken has excellent research resources for scholars and many kinds of rare electrostatic apparatus on display.

The Burndy Library, Electra Square, Norwalk, CT 06856. Houses and publishes books on the history of electricity. Excellent sources for scholars.

Electret Scientific Company, P.O. Box 4132, Star City, WV 26505. Publishes books on electrostatics applications, electret waxes, electrostatic motors and experiments.

Electrostatics Society of America, c/o Dr. Emery P. Miller, 641 East 80th St., Indianapolis, IN 46240. The E.S.A., founded by Dr. A. D. Moore, presents awards for outstanding science fair projects, holds yearly conferences and publishes a newsletter. Open to anyone interested in the study of electrostatics.

Energy Unlimited Publications, P.O. Box 493, Magdalena, NM 87825. Publishes *Causes Newsletter*.

Index

A

abstract thought/reasoning, 179-183
acupuncture, electro-, 130
Aghandlung von dem Luftelektrophor, 103
Andrew, Gustave, 151
Annals of Electricity, Magnetism and Chemistry, 133
anomalous electrical discharge, 163-178
  auroral displays, 137-138
  ball lightning experiment, A.T. Jones, 173
  ball lightning experiment, Arden and Constable, 1, 174
  ball lightning experiment, Weisiger, 171-173
  ball-shaped, eyewitness accounts of, 168
  countergravitation and, 155-156
  damage caused by, 176
  earthquake lights, 90, 140
  electrical entities, 168-178
  experiments with, historic examples, 169-176
  formation of, conditions conducive to, 168-169
  gravitation disturbance,countergravitation, 177
  Lifesaver-candy discharge, 140
  lightning shadowgraphs, 163, 164
  negative entropy and, 178
  phosphorescence, 141-142
  physics of, 178
  rarified gases, 142-143
  theories concerning cause and appearance of, 176
  thermodynamics and, 178
  tornadoes as electrical generator, 57, 91, 163, 165-168, 170
  triboluminescence, 140
Apostoli, 56
applied voltage, 94
Arden, 173, 178
artificial-gravity field generator, 162
atmospheric electrostatic motors, 120
auroral displays, 137-139

B

Bacquerel, Edmond, 142
Baker, Burton, 59
Ball Lightning and Bead Lightning, 176
Barry, James D., 176
Becker, Robert, 133
belt tensioning, induction generator, 26
Bennet, Abraham, 11
Bertholon, Abbe, 169
Birkeland, Kristian, 138-139
Body Electric, The, 133
boilides (*see* anomalous electrical discharge)
Bonetti Wimshurst-generator, 17
bosses, induction generator disc supports, 22-24
brain trusts, 182-183
Brand, Walther, 168, 176
Brown, Thomas T., 162
Brush, Charles F., 156
Buchanan, Relva, 166

C

Cahn, Maurice, 151
Canton, John, 11
capacitors, spontaneous self-motion of, 162
carriers, 17
Cavallo, Tiberius, 106, 169
Ceramics Research Laboratory, 166
charge collector, induction generator
  combs for, 28-34, 72
  support system for, 26-28
charging, induction generator, 43-46, 74
Chaumat generator, 56
Chaumat, Henri, 56, 59
cold light anomalies, 137-144
combs, charge collector, induction generator, 28-34, 72
Complete Treatise on Electricity in Theory and Practice, A, 169
condenser, Leyden jar as, induction generator (*see also* Leyden jars), 46-51
Constable, William, 173, 178
Corliss, William R., 91, 140, 167
corona discharge, 74, 151
countergravitation, 153-162, 177
  artificial-gravity field generator, 162
  kinetic gravitational theory and, 154-156
  levitation, 158-161
  Nipher's deflection experiments, 156-158
  Piggott's levitation experiments and, 158-161
  Wave Theory of Gravity, 162
Crookes, William, 87, 142
current draw, induction generator, 21
cylindrical generator, 57

D

D'Arsonval, 56
Dailey, Howard B., 117
De L'electricite des Meteores, 169
deflection and countergravitation, 156-158

Delaval, Edward, 98
Dempster, James, 59
Der Kugelblitz, 168
design, 5-9
diathermy, 130
Dibner, Bern, 3
dielectric constants, insulators and Leyden jars and, 94-96
dielectric plate, electrophorus, 107
dimetral conductors/inductors (*see* neutralizers)
Dirac, Paul, 154
discharge terminals, induction generator,
  spacing of, 74
  supports and rods for, 34-39
discs, induction generator, 19
  bosses for, 22-24
  materials used in, 59, 66, 79
  support shaft for, 22
dissociation of matter, sunlight exposure and, 89-90
doubler generator, 11
drive belts, induction generator, 24-26
drum generator, 57
Dudley, Horace C., 151-152
dust-impact generator, 58-65
Dwelshauvers-Dery, F.V., 73
dynamic theory, 91

E

Early Electrical Machines, 3
earthquake lights, 137, 140
  electroscope, spontaneous charge in, 90
Ebert, Hermann, 61, 142
electrets, 105-106
"electric egg", 142-143
electrical entities, 168-178
Electrical Experimenter, 157-158, 169
electrical fields, 145-147
Electrical Influence Machines, 70
electrical meteors (*see* anomalous electrical discharge)
Electrical Review, 53-56
Electrician, The, 18
Electricity in Agriculture and Horticulture, 123
Electricity in the 17th and 18th Centuries, 94
electrification, 11, 74, 76, 112, 114, 115
electroacpuncture, 130
electroaerodynamics, 151-152
electroculture (*see* electrohorticulture)
electroencephalograph (EEG), 130
electrohorticulture, 123-126
  Lemstrom's generator for, 124
  research results from experiments with, 124-126
electrolysis, 130
electrophorus, 76, 101-115
  charge indicator, 110-112
  charge-radiation from, 112-113
  charging of, 102
  construction steps for, 104-112
  dielectric plate in, 107
  discharge capacity, electric shock from paper, 103-104
  grounding, 103
  heat source for, 108-109
  metal cover and insulating handle for, 109
  operational theory of, 101-103
  parts of, 101
  research avenues for, 114
  resin cake for, 104-106
  sealing wax for, 106-107
  semiconductive stones and, 112-114
electroscope, 76-91, 115
  anomalies of, 87-89
  construction of, 81-86
  current sensitivity of, 84, 86
  dissociation of matter, sunlight exposure and, 89-90
  earthquakes, spontaneous charging in, 90
  energy "nodes" around, 87
  Faraday cylinder in, 81
  "magic lantern" for display of, 86
  operation of, 78-80
  parts of, 77
  radioactive ionization, experiment with, 88
  research avenues using, 89-91
  schematic of, 82
  three-state experiment, 88
  tornadoes, spontaneous charge in, 91
electroshock therapy, 130
electrostatic generator, 57, 58
  frictional-type, 3-4
  induction-type, 3-4
Electrostatic Motors, 117
electrostatic motors, 117-121
  atmospheric type, 120
  construction of, 118-121
  cylindrical-type, 118
  disc-type, 118
  early experiments with, 117-118
  grounding, 118
  operation of, 120-121
  static electric top, 119-120
electrotherapeutics, 127-131
  advertisement for, circa 1905, 129
  early experiments in, 127-130
  electroencephalograph (EEG) and, 130
  electrotherapy generator, circa 1908, 128
  modern applications of , 130-131
  usage of, illustration of, circa 1908, 128
Elektrotechnische Zeitschrift, 57
English Mechanic, 5, 163
entities, electrical, 168-178
Essentials of Modern Electro-Therapeutics, 128
Evolution of Forces, The, 87, 141
Evolution of Matter, The, 87-88

F

Faraday cage, 88
Faraday cylinders, 81
Felici generator, 57, 58
Felici, Noel, 57, 59, 67
Ferguson, Robert M., 5, 6, 8
Fessenden, Reginald, 126
fireballs (*see also* anomalous electrical discharge), 140, 166-168
Frick, J., 8
frictional dust generator, 58-65
  dust material for, 63-65
  limitations, theoretical, 62-63
  operational principle of, 60-66
  performance chart, 65
frictional generator, 3-9, 58, 178
  operation of, 7-9
  parts of, 6, 8-9
frictional slippage, 3, 140

G

gas chambers, 67, 74
Gee Sun, Chang, 133
Geissler, 56
General Electric Review, 178
Golden, Kenneth, 126
gravitational disturbances (*see* countergravitation)
gravitons, 154
gravity, effect of electrification on (*see* countergravitation)
Gray, John, 70

grounded electrophorus, 103
grounding, 14

H

hair-raising experiment, 145-146
Harris, William Snow, 104
Hart, W., 8
Heilbron, J.L., 94
Helmholtz, 62
Helpel, 67
Hertz, Heinrich, anomalous electrical discharge theory, 177
Hoffman, 61
Holtz generator, 11, 54, 56, 74
Hooper, William J., 162
horticulture, electro-, 123-126
human body, high-voltage generation, 133-136

I

impact generator, 58
impact, charge created by, 3
incandescent light, 137
induction, 3, 69
induction generator, 3-5, 19-51
  belt tension in, 26
  charge collector, combs for, 28-34, 72
  charge collector, support system, 26-28
  charging the generator, 43-46
  current draw, 21
  dimensions of, overall, 21
  disc, size variations in, 19
  discharge terminals,supports and rods for, 34-39
  discs, bosses for, 22-24
  discs, support shaft for, 22
  drive belts and pulleys for, 24-26
  Leyden jar condensers in, 19, 46-51, 72
  motor for, 20, 21
  motors, mount for, 26
  neutralizer rods and blades for, 39-41, 44-45
  operational theory of, 69-74
  ozone production, ventilation for, 46
  power supply for, 20, 41-42
  sectorless Wimshurst-design, 20, 72, 76
  spark length in, 19
  stand for, 20-22
  Wimshurst-type (*see* Wimshurst generator)
inductors, 67
influence machine (*see* induction generator)
influence, theory of, 69-73
Ingenhousz, Jan, 3
insects, light-producing, 137
insulators, dielectric constants for, 94
Introduction to Experimental Physics, 78
intuition, invention through, 180
invention, thought processes of, 179-183

J

Jefferson, Thomas, circa 1920, 162
Jefimenko, Oleg, 117
Johnsen-Rahbek effect, 113
Jolivet generator, 56
Jolivet, Pierre, 56, 59
Jones, A.T., 169, 173
Journal of the Rontgen Society, 54, 129

K

kinetic gravitational theory, 154-156

L

Lazarus-Barlow, W.S., 88
Le Bon, Gustave, 87-90, 141, 142, 147-149, 177
Le Sage, Georges Louis, 154
Leduc, Stephane, 169, 171
Lemstrom generator, 18, 124
Lemstrom, Selim, 123, 137
levitation, 146-149, 158-161
Lexan plastic, 67
Leyden jars, 11, 14, 16, 19, 56, 72, 76, 93-99, 114-115, 127, 145, 173, 176, 178
  batteries of, joining multiple, 95, 97
  condenser used in induction generator, 46-51
  design modifications for, 96-99
  discharge capacity of, shock hazard, 98
  discharging, discharge tongs for, 99
  grounding of, 93
  insulators, dielectric constants and, 94-96
  invention of, 93, 94
  jar selection for, 95
  lids for, 95
  liquid condenser type, 97
  physics of, 98
Leyser generator, 11
Lichtenberg, 105
Lifesaver candy, electrical discharge from, 140
lightning balls (*see* anomalous electrical discharge)
lightning discharge, 58
lightning shadowgraphs, 163-164
Lightning, Auroras and Nocturnal Lights, 91, 140
liquid condenser, Leyden jars, 97
liquid, gas, vacuum chambers, 67, 74

M

Maxwell, James Clark, 178
mechanistic theory, 91
medical uses of electricity (*see* electrotherapeutics)
microwave heating, 130
motors, induction generator, 20-21, 26
"mouth lightning", 140

N

Nature, 177
negative entropy, 178
neutralizers, 39-41, 44-45, 56, 59, 74
neutrino flux, 154
New China News Agency, 133
New York Times, 133
Newton, Isaac, 153-154
Nipher, Francis, deflection and countergravitation, 156-158
Northrop Norair Division, electroaerodynamics, 151
nuclear magnetic resonance, 130

O

On the Cause of Magnetic Storms, 138-139
Oudin coils, 56, 128
ozone production, induction generator, 46

P

pacemakers, 130
particle physics, 154
perpetual electrophorus, 101, 102
Phillips, J., 103
Philosophical Magazine, 103
Philosophical Transactions, 112
phosphorescence, 141-144
phosphorescent lamps, 143-144
photography
  invisible phosphorescence and, 141-142
  lightning shadowgraphs, 163-164

Physical Review, 60
Physidalische Technik, 86, 93
Picolet Wimshurst-generator, 17
piezoelectric discharge, 90
Piggott, George, countergravitation and levitation, 158-161
plants, effect of electricity upon (*see* electrohorticulture)
plastics, 67
Poincare, Lucien, 178
polarity reversal, 12, 74
Popular Electricity, 58, 97
Popular Mechanics, 112
potential difference, 74, 94, 145-146
power supplies, induction generator, 20, 41-42
Practical Technics, 8
precision construction techniques, 67
Preston, Thomas, 178
Principia, 153
Product Engineering, 152
pulleys, induction generator, 24-26
Puluj, Johann, 142-144

Q

qualitative anatomization, invention through, 180-182
quantum physics, 154

R

Radio-activity, 84-85
Ramsay, William, 88
Ramsden, Jesse, 3
rotors, 67
Rudge, W.A. Douglas, 61
Rutherford, Ernest, 84-85, 115

S

Samen, E.H., 58
Schaffers Wimshurst-generator, 18
Science, 138, 169
Science Abstracts, 73
Science and Invention, 133
Scientific American, 17, 163
scientific thought, 179-183
Second Law of Thermodynamics, 178
sectorless Wimshurst machine, 17, 20, 72, 76
Selden, Gary, 133
self-starting generator, 74
semiconductive stones, electrophorus and, 112-113
shadowgraphs, lightning, 163, 164
soft particles, 154
spark length, induction generator, 19
St. Elmo's fire, 137
Stager, A., 61
stand, induction generator, 20-22
static electric top, 119-120
static generator, 58
statics, 91, 111
Steinmetz, Charles P., 178
Sweeting, Linda, 140

T

Tesla coils, 123, 128
Tesla, Nikola, 56, 180
Thermodynamics, Second Law of, 178
think tanks, 182-183
Thomson, Elihu, 173
Todd, Henry, 59
Toepler generator, 11
top, static-electric powered, 119-120
Tornadoes, Dark Days, Anomalous Precipitation and Related Weather, 167
tornadoes, electrical generation by, 57, 91, 163, 165-168, 170
Transactions of The Academy of Science of St. Louis, 157-158
triboelectric discharge, 3, 90
triboluminescence, 140
Tributsch, Helmut, 90
Tudsbury, 67

U

ultramundane corpuscles, 154

V

vacuum chambers, 67, 74
Van De Graaf, Robert, 67
Van de Graaff generator, 58
van Musschenbroeck, Pieter, 93
Varley generator, 11, 74
Vassalli-Eandi, 90
virtual particles, 154
Volta's Paradox, 113
Volta, Alessandro, 101, 112
von Guericke, Otto, 3
von Humboldt, Alexander, 90
von Kleist, Ewald, 93

W

Wave Theory of Gravity, 162
Weather, 166, 170
Weber, Joseph, 103
Weinhol, Adolf, 78
Weisiger, Samuel, 169, 171-173
When the Snakes Awake—Animals and Earthquake Prediction, 90
Wilcke, Johnannes, 101
Wilson, C.T.R., 84, 87
Wimshurst generator, 11-18, 54, 56, 59, 67, 72, 114, 141, 161
  construction of, 12-16
  glass disc varnishing,process for, 15
  largest built (1885), 13
  Leyden jar condensers in, 14, 16
  modifications and improvements to, 16-18
  multi-plate type, 14
  operational theory of, 69-73
  parts of, 12
  polarity reversal, 74
  sectorless, 17
Wimshurst, James, 11-18
windmills, horsepower developed by, 173
Winter, Karl, 3, 5-9, 59
Wommelsdorf generator, 17, 53-57
  applications for, 55
  improvements in design of, 56-58
  Leyden jar condenser in, 56
  operation of, 54-56
  parts and construction of, 54
  performance chart, 55-56
Wommelsdorf, H., 54

Z

zero-point energy, 154